THE HOMEOWNER'S ILLUSTRATED GUIDE TO CONCRETE

BY DAVID H. JACOBS, JR.

TAB BOOKS Inc.
BLUE RIDGE SUMMIT, PA. 17214

FIRST EDITION

FIRST PRINTING

Library of Congress Cataloging in Publication Data

Jacobs, David H.
The homeowner's illustrated guide to concrete.

Includes index.
1. Concrete construction—Amateurs' manuals. I. Title.
TA682.42.J33   1984   693'.521   83-24214
ISBN 0-8306-0626-2
ISBN 0-8306-1626-8 (pbk.)

Contents

Acknowledgments

THANKS TO R. KEITH REIBER AND GERALD BREWster for taking photographs and to Lloyd Smith for processing film. Special thanks to Dave Reum and the men of Custom Concrete and to Southern Oregon Concrete.

Introduction

IN THE LAST FEW YEARS, CONCRETE HAS BE-
come so expensive that some refer to it as "gray
gold." Because of that cost and the labor fees, many
homeowners are going without concrete walkways
and patios. If your concrete job would cost only half
as much as the estimate, I'm sure you would get it
done right away.

I have poured hundreds of yards of concrete for
homeowners for several years. Having done a lot of
work in the surburbs of San Diego and the rural
areas of Southern Oregon, I am familiar with the
needs and wants of most homeowners. Basically,
they are all the same—walkway down the side of
the house and a patio in the backyard. These are
rather simple jobs that can easily be done by the
do-it-yourself homeowner.

I can't count the number of times homeowners
have told me that they were going to do the job
themselves, but . . .. Their "buts" were, for the
most part, simple problems with easy solutions. I
also realized that almost 50 percent of the cost for
most of the jobs I did went for my labor.

I have had a difficult time locating a book that
describes the basics of concrete work—one that
stays strictly on the subject of concrete flatwork:
patios, walkways, driveways, and mowing strips.
Many books about masonry and brickwork do include a chapter or two on concrete flatwork, but
they do not go into great detail on custom ideas,
concrete yardage calculations, ordering delivery,
pouring, and finishing.

This is where this book has the others beat.
From page 1, I take you through each and every
phase. From designing your job to the point when
you can stand back and admire your work, everything is covered, described, and explained. You will
learn how to design a slab or walkway that conforms
with the landscape in your backyard, how to properly form it, determine how much concrete it will
need, the cost, the amount of help you'll need, how
to pour and finish it, how to protect it, and how to
apply custom finishes.

I don't claim that you will be a qualified concrete professional after reading this book. Like ev-

erything else, you can't learn all there is to know by simply reading one book. Therefore, I have included ways for you to get additional information and help in your own area. In many cases, a simple phone call to the right person will get your question answered and the problem handled. If you are not sure about a particular phase of your concrete work, never fail to ask for help or advice. As a professional concrete man, I have never avoided the opportunity to help out a novice. I am sure that any other concrete man would be happy to help you. In some situations, he may charge a small fee for his time and inconvenience. In the long run, though, the few dollars it may cost for advice will save you plenty when the problem is correctly abated.

The Homeowner's Illustrated Guide to Concrete was written for you, the homeowner. Following the instructions, advice, and recommendations will give you the opportunity to form, pour, and finish your own concrete work. Along with the self satisfaction, you'll also save a lot of money.

Chapter 1

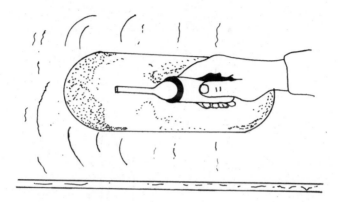

Design Considerations

IN CONCRETE WORK, TAKE AN AMPLE AMOUNT of time to plan exactly what type of job you want. Get a chair, a pencil, some paper, and something cold to drink. Sit near the proposed site and start drawing plans (Fig. 1-1). The blueprints do not have to be perfect or to scale—just usable and easy to read. Sketch a few different plans. Add or delete curves, stringers, and other custom features. Always remember the layout or future design of the existing landscape and try to blend the concrete work with it.

WALKWAY

Decisions must be made as to variables. For example, if you are going to pour a walkway down the side of your house, do you want the concrete to butt next to it or would a planter be more presentable (Fig. 1-2)? Leaving space for plants or shrubs breaks up the solid concrete look and makes for a professional looking landscape.

Factors that may aid in making such a choice would be whether or not that side of the house gets enough sun to support plants, if visitors will see and use the walk, or if it will be used to store tools and trash cans. Placing the concrete next to your house might be attractive and give you more room in the yard (Fig. 1-3).

Curves

Another walkway facet is curves or straight runs and corners (Fig. 1-4). If your yard is basically landscaped with straight planters, you could angle the walk for variation or design one that incorporates both right angles and curves (Fig. 1-5). Try to maintain an even blend between curves and angles (Fig. 1-6). For example, use the curved feature on the outside and keep the inside angles at 90 degrees.

Some walkways can be *free-formed*—that is, to form it using many curves winding through the yard (Fig. 1-7). Designs can be determined by trees or banks (Fig. 1-8). These walks can also be dressed up by adding a brick border after the concrete has cured (Fig. 1-9). Free-formed walkways and those

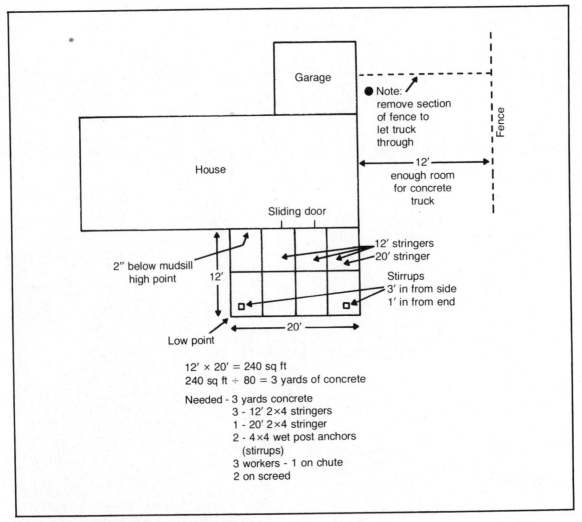

Fig. 1-1. A rough blueprint for your concrete job.

with rounded corners will help the job better fit the rest of a circular landscape. Forming these types of jobs is simple, and I will clearly show you how in Chapter 10.

Width

The width of your walkway is also an important consideration (Fig. 1-10). Most walks are 3 feet wide, making for ample space to push a wheelbarrow or lawn mower across. Four feet or more is uncommon, but this width may be perfect for your needs (Fig. 1-11). A wide walkway is really a minislab and can be useful for settees or as a nice front porch (Fig. 1-12). The added width can be broken up with stringers or brick inserts. The dimensions are controlled only by your imagination and needs.

A narrow walkway less than 3 feet is functional for only one person to walk on (Fig. 1-13). This type would be adequate in areas of little use and mainly for decorative purposes.

If you intend to lay a 3-foot-wide walk, form it at 3 feet 1 inch (Fig. 1-14). The added inch will not make much difference in appearance, but it will be very handy when finishing time comes. Most all large concrete tools such as a tamp, bull float, and fresno are 3 feet wide. The extra inch in width will accommodate those tools and allow you to use them down the length of the walk rather than across 3 feet at a time.

For example, if you form up a walk 3 feet wide and 20 feet long, it will be much quicker to push the tools down the 20-foot run in one action than to have to finish the concrete across the 3-foot width back and forth, covering only the 3-foot area covered by the tools.

Slope

Water runoff is a very essential factor. Ideally, the concrete will run downhill toward the street or yard (Fig. 1-15). This will force any water accumulation to flow away from the house. As you are sketching plans, determine where you'll want the water runoff to go and design the walk in that way. Water from

Fig. 1-2. Planter between the house and walkway.

Fig. 1-3. Walkway against the house.

rain and washing off the concrete will go where you want it if the job is formed and placed correctly. Under no circumstances will you ever want water to run toward your house. Always be certain the concrete slightly slopes toward the yard with the highest point next to the house (Fig. 1-16).

Generally, a good guide to show slope will be in the mudsill on your house (Fig. 1-17). That is the lip made when the stucco wall meets the foundation. It will be very close to level. Use it to show a high point while forming the walk to run downhill. The space between the top of the concrete and the mudsill will be less at the high point and much greater at the low point. Water will run toward the low point.

I have had to break out other people's concrete work and repour the entire job because they did not consider water runoff. Water puddled next to the house or in low spots on the yard makes life miserable during the winter. All of the hassle and extra

Fig. 1-4. Front porch and walkway with straight runs.

spray some water on the roof and measure how far from the house it drains off. Form the concrete accordingly (Fig. 1-19).

PATIO SLABS

Some of the ideas mentioned about walkways are also true for patio slabs. The corners serve no special function except for looks (Fig. 1-20). The corners can just be rounded, or each side can be free-formed. Some designs incorporate both the free-form characteristic and planters (Fig. 1-21).

Other features can include extensions for cover posts, which give you more room on the slab (Fig. 1-22). Also, you can pour two separate patios of different dimensions connected by a short walkway (Fig. 1-23). Any design is good as long as it is functional and you like it.

Slope and depth are two factors that can go hand in hand (Fig. 1-24). No slab you pour needs to be deeper than 4 inches. Normally 2 × 4 forms are used, and the thickness of the concrete is the same as the form. When determining the high point, take a good look at the natural slope. Sometimes it is easier to pour the slab with the natural grade (Fig. 1-25). If the grade is not proper, you will have to remove dirt from the high side and fill in the low, dig out only enough to make room for a 4-inch slab (Fig. 1-26). Unnecessary digging does no good and makes the job longer and harder.

The use of 2×4 screed boards will help you determine the grade, especially if you use 2 × 4 stringers in the slab. The only jobs that require deeper slabs are specialized pads for heavy commercial use and certain footings. Those applications will be covered later.

CUSTOM IDEAS

Most people that I have done work for like their concrete "broken up." By that, I mean doing something different than just an old common slab. Custom work is easily done without any special problems. Sometimes the pouring even gets to be fun.

Stringers

The use of *stringers* is one way to customize a slab

work would have been avoided if the concrete former had done his homework and specifically designed the job for runoff.

Let's not forget about rainwater running off the roof, onto the ground, and splashing mud against the house. There are two ways to solve this problem. You can invest in rain gutters. Use caution as to where they drain (Fig. 1-18). An open-ended gutter can quickly flood a flower bed. Use 3-inch drainpipe under the walk and direct flow toward the street.

Another way to avoid mud splatter is to place the inside edge of the concrete close enough to the house so that rain drips will fall on the concrete first. Make the width of the planter less than the roof overhang. To assure yourself of that distance,

5

Fig. 1-5. Walkway with right angles on the inside and curves on the outside.

(Fig. 1-27). Redwood or pressure-treated fir in 2 × 4 lengths are placed inside the forms, and the concrete is poured against them. Some areas can be left open for planters, decorative bark, or colored rock.

The stringers are placed on the 2-inch side, with concrete placed against the 4-inch sides. The end result will be the showing through of the top 2-inch side (Fig. 1-28). These tops can be stained to match other exposed wood around the house. You might even want to leave the entire sides showing on steps or planters (Fig. 1-29). Bracing and anchoring stringers is critical (see Chapter 12).

Stringers can be placed and designed any way you want. Most jobs have them running perpendicular to each other. They can evenly divide the slab, or they can produce many rectangles—none of which are the same size. Also, 1 × 4 wood can be used (Fig. 1-30). The only drawback to this smaller

Fig. 1-6. Curved walkway joining a rectangular patio.

width is that it is more difficult to keep in place, and it bows much easier than the 2-inch wood.

Stringers take more time to place and will add to the cost of your job, but the custom appearance might just be worth it. Designs for stringers are unlimited. You can use almost anything for a stringer, including railroad ties (Fig. 1-31).

Broom Finish

The type of finish you choose for your slab can really enhance the final outcome. There are many finishes that you can apply, with the broom finish being the most common (Fig. 1-32).

Most driveways and sidewalks have a broom finish. It is the easiest to apply and can cover many finishing errors. Small lines left by the trowels will be wiped away by the broom leaving behind a rather smooth slab with an even texture. A common horse-hair push broom with soft brushes is used.

The broom finish also can be customized (Fig. 1-33). The broom can be applied to wet concrete, and the edges can be gone over again with the edging tool. The end result will be an outline of smooth concrete around a rather rough center.

Traction on wet concrete is important. We all know that when wet, concrete can be as slippery as ice. On all outdoor slabs and walkways, always put on some kind of finish. The broom finish will keep the concrete from becoming slippery and greatly add to traction (Fig. 1-34).

7

Fig. 1-7. A free-formed walkway.

Color

I'm sure you have seen porches and walkways done in colored concrete. It is very pretty and is applied in one of two ways. You can order the entire load of concrete colored from the batch plant, or you can purchase the powdered dye and spread it on the surface yourself (Fig. 1-35).

I prefer ordering the concrete colored from the plant. First, it is easier to work with because there will be no dye to apply. Second, the entire load of concrete will be the same color rather than only the top fraction of an inch. If a corner of the slab should crack off at a later date, the material underneath would be gray on a hand-dyed job, whereas the cracked part would match on one poured all in one color.

Generally, finishers apply a special wax or sealer over the surface of colored concrete. This will protect the dye and keep the slab shiny. If you decide to use colored concrete, check with the concrete dispatcher and request his assistance with choosing a color and sealer. He can also advise you on appropriate measures to be taken with reference to your particular climatic area. Locations with harsh and freezing winters pour a slightly different mix of concrete than those with mild winters. Your concrete dispatcher will be aware of these problems and will be more than happy to help you.

Traction

As mentioned earlier, the broom finish will provide a degree of traction to wet concrete (Fig. 1-36). Other finishes will also help. The rock salt finish is usually lightly broomed before the salt is spread

Fig. 1-8. Free-form walkway with curves at trees.

Fig. 1-9. Curved walkway dressed with a brick border.

Fig. 1-10. The width of your walkway is an important factor.

(Fig. 1-37). This will ensure that every part of the slab has been textured to reduce slippery results. The etching designed slab is also broomed for traction before the pattern is made (Fig. 1-38). The only type of finish that doesn't need to be broomed is the exposed aggregate (Fig. 1-39). By itself, it provides ample traction even when wet.

Almost as important as slope, traction must be a prime consideration. Wet concrete can become a dangerous hazard during rainy weather. Many people have slipped on that wet surface and have become extensively injured. Be sure to include some type of traction promoting finish to your concrete work.

Rock Salt

The rock salt finish is a pretty one, especially for

Fig. 1-11. A wide walkway will allow room for a bench.

patios (Fig. 1-40). It is a finish that has hundreds of tiny, shallow holes in the surface. The finisher will trowel the surface smooth, apply a very light broom, and then evenly spread handfuls of rock salt over the entire surface. He will get out on the slab and pound in the salt with a hand float or trowel. After the salt has dissolved, the surface will be textured with hundreds of small holes (Fig. 1-41).

The size of granules you use is important. I have been satisfied with the coarse brand. The granules are generally sized between the range of three-fourths of a pea to the same size as a pea (Fig. 1-42). Smaller granules are available as well as larger ones. Your preference is really the only guide. The main principle to remember with rock salt is to maintain an even coverage. Areas spread too heavy will look awkward next to those too thin.

Although the rock salt finish is eye appealing, it is more difficult to keep clean (Fig. 1-43). A simple brooming will not remove dust and dirt from the tiny holes. You usually have to wash down the slab with water to get it clean.

Fig. 1-12. Wide walkway along the front of a house. It can also be used as a front porch.

Exposed Aggregate

A unique finish is the exposed aggregate, and it is truly custom (Fig. 1-44). The finish is made by simply washing off the top layer of "cream" to expose the rock (aggregate) underneath. It looks very nice, blends in well with landscapes, and is easy to create (Fig. 1-45).

Like rock salt finishes, it is harder to keep clean. The rough surface makes sweeping difficult and will require that you wash it off more often with water (Fig. 1-46). This finish is also very hard on bare feet and especially hard on the knees and elbows of youngsters who might fall down on it.

One way to avoid some of the problems with the rough texture is to order smooth rock in the concrete. Some concrete companies get their

Fig. 1-13. Narrow walkway for a parklike path around the yard.

aggregate from river bottoms where the gravel has been worn smooth by water. It is sometimes scarce, and you might have to ask for it by special order. Again, your concrete dispatcher will be glad to assist you if he can.

Aggregate comes in different sizes, and you can specify which size you want (Fig. 1-47). Most finishers use ¾-inch gravel, although many now are using the much smaller pea gravel for a finer look. The size you choose will be determined only by the appearance desired.

Producing an exposed aggregate finish will require washing off the very top layer of cream from the entire job. It can get very messy. The cement

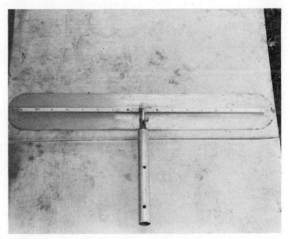

Fig. 1-14. A walkway 3 feet, 1 inch wide accommodates 3-foot-wide tools.

11

Fig. 1-15. The slope of the patio should be away from the house.

powder, sand, and water will have to go somewhere. Be prepared for a messy cleanup job afterward.

Etching

One of the newer finishes to be used in recent years is simply called the etching finish (Fig. 1-48). It is easy to apply and also a lot of fun. There are no set patterns; it is mostly all free-form. The only parts that you will have to keep the same are the intersections; they must all be rounded out.

The concrete is finished with the trowel, broomed, and then etched with a 1-inch-wide wire brush (Fig. 1-49). The wire brush does not have to go very deep—only enough to leave the impression (Fig. 1-50). If you want, you can go deeper and expose the aggregate underneath. On a job that you tamped, you might have to go too deep, and the etching might not be uniform.

All edges and expansion joints will be etched, and the corners should be rounded (Fig. 1-51). Then you can start to free-form. The design will

12

Fig. 1-16. Against the house the level is flat; the slope runs toward the backyard.

resemble a flagstone deck, with the etched lines appearing like the mortar between the steps. The etched lines will also remain a darker color than the rest of the slab, adding to the contrast and giving the appearance of a special slab (Fig. 1-52).

Not too many of these etched designs have been done. Most that I have seen have been done by me or some of my friends. I'm sure that this design is catching on, as all of my customers are very happy with it and have told others about it as well.

Inserts

Many concrete finishers have used brick to liven up

Fig. 1-17. Use a mudsill lip on the wall as a level guide. Notice that the slab is level with the back wall, but it slopes against the wall on the left.

Fig. 1-18. Rain gutter drain under the walkway.

their slabs (Fig. 1-53). Walkways can be bordered by red brick to extend their width and match existing brickwork around the house (Fig. 1-54). Patios can be bordered and sectioned by brick to create a warmer feeling (Fig. 1-55). Brick can also be used to make special designs in the middle of slabs, which can later outline a special planter box or decorative statue (Fig. 1-56).

Brick inserts are made after the concrete has been poured. The areas where the brick will eventually fit should be formed up as though no concrete will go into them. After the concrete has cured, you can place the brick and mortar (Fig. 1-57).

Decorative Seams

Similar to stringers, decorative seams give eye appeal to a slab and help to get rid of the solid look (Fig. 1-58). They are applied with a special seamer tool and require the use of a straight 2 × 4 board as a

Fig. 1-19. Narrow width of the planter allows rainwater runoff to hit concrete and then roll into the flower bed.

guide. The lines can be made at angles, or they can be matched to create a slab composed of several equal rectangles (Fig. 1-59).

Using decorative seams along with a special finish and planters can add much to an ordinary patio. There is more work involved, but the everlasting quality will remain forever.

Seams must be initially made after the bull float application while the concrete is still wet. This will ensure that no rocks will be in the way later when the final seams are made. In the case of a large patio, it would be wise to use a seamer that has the ability to accept extensions. This will enable you to reach all areas of the slab without having to get out on it until the time for final finishing (Fig. 1-60).

PATIO PLANTERS

When doing jobs with redwood stringers, often a few of the squared areas are left open for planters. This idea has worked great for many jobs. With a

14

Fig. 1-20. Free-form curved sides on a patio.

little foresight, you can incorporate a small planter into your slab (Fig. 1-61).

Planters can be used as partitions, screens from excessive afternoon sun, or to create a secluded section of patio for private use. Shrubs, hedges, and trellises fit well into such planters and add a very custom look to your work.

MOWING STRIPS

Most people like a well-manicured yard and don't appreciate unwanted grass growing in the flower beds. The solution may lie in mowing strips (Fig. 1-62).

Mowing strips are thin lengths of concrete that separate special areas. They can be used to divide gravel driveways from the lawn, or they can simply border the base of a tree (Fig. 1-63). The depth varies. For the most part, 1 inch under the ground and 3 inches above are sufficient. If you have Bermuda grass, you might want the strip to go much

Fig. 1-21. Free-form patio with planters on the inside.

deeper into the dirt. These concrete borders can separate anything from grass to colored rock to different types of foliage. They can also be designed to match the landscape, whether it be rectangular or circular (Fig. 1-64).

Actual pouring of mowing strips is tedious. The concrete must be placed by shovel, because the area is not wide enough to accommodate a wheelbarrow. Not much concrete is needed for them either, because most strips are only 4 inches wide. A good way to provide concrete for this project is with the 1-yard hauler (described later) or by using any concrete left over from another job.

EXPANSION JOINTS

An *expansion joint* is created when a piece of special felt or redwood benderboard is placed in the concrete to absorb the expansion and contraction of concrete as it heats and cools (Fig. 1-65). Especially in walkways and at points where a slab meets another piece of concrete, expansion joints prevent cracks (Fig. 1-66).

These joints should be made every 10 feet on a walkway and at every point where there is a corner or intersection of concrete. If a walkway extends from a patio slab, there should be a joint at that intersection. If not, the concrete will surely crack there. Remember and plan for adequate joints. The material you'll need is available at local lumberyards and from the concrete company. Note that any time you join your concrete with a city

sidewalk, you will be required to place a piece of at least ½-inch-thick felt between your concrete and the sidewalk.

SPECIAL BORDERS

Concrete can outline or border a number of things and it can also become a border itself. Besides a border of brick, certain stepping stone tiles can add an element of grace to a slab (Fig. 1-67). The concrete can also become its own border with some special and customized edging. In Fig. 1-68, the finisher simply used his wide walking edger backward. Instead of placing the arc portion of the tool next to the form, he allowed it to scribe the surface. He gently used a seamer tool to complete the seam. The effect appears like a curb surrounding the slab.

On a specialized note for those of you with asphalt driveways, one way to prevent the end of the drive from breaking off, due to auto use, is to build a concrete ramp (Fig. 1-69). First, use a concrete cutter to square off the asphalt. Then pour the concrete against it just like any normal slab. This will ensure against asphalt breakage and make a cleaner looking apron.

Fig. 1-22. Small slab extensions for patio post supports.

Fig. 1-23. Unique design of one slab with three different dimensions—almost like three separate slabs.

Fig. 1-24. Depth of the slab should be no more than 4 inches. Slope should be away from the house ¼ inch per foot.

Fig. 1-26. The depth of the concrete can be as little as 3½ inches—the height of a 2 × 4 form.

Fig. 1-25. Use the natural grade as a slope guide. Try to go with it if possible.

Fig. 1-27. A patio slab using a number of stringers. Note the design areas left open for the decorative bark.

Fig. 1-28. Only the top side of the stringer will show.

18

Fig. 1-29. Stringer step face forms left in place. All stringers are stained to match the woodwork on the house.

Fig. 1-30. Walkway using 1 × 4 stringers.

Fig. 1-31. Custom walkway with railroad ties as stringers.

Fig. 1-32. Broom finish on the driveway. Note planters on each side of the garage door.

Fig. 1-33. Custom broom finish. Edger tool was used after the broom to smooth the outer edges.

Fig. 1-34. Extra traction is needed for concrete around planters that, when watered, will cause the concrete to get wet.

Fig. 1-35. Color dust applied to curbing design in a patio.

Fig. 1-36. A light broom finish for a small patio.

Fig. 1-37. The concrete slab was lightly broomed before the salt was applied.

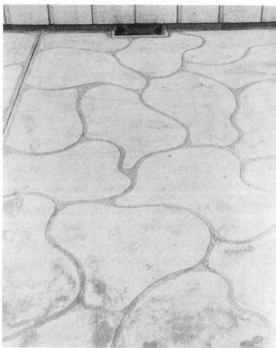

Fig. 1-38. The slab was lightly broomed before etching.

Fig. 1-39. A broom finish is not necessary on exposed aggregate.

23

Fig. 1-40. The rock salt finish has a lot of eye appeal.

Fig. 1-41. Texture of the concrete after a rock salt finish.

Fig. 1-42. Rock salt granules should be no bigger than a pea.

Fig. 1-43. A salted slab must be hosed down more often to keep it clean.

25

Fig. 1-44. Very eye appealing walk-way with brick inserts.

Fig. 1-45. Walkway with offset porch and planter looks nice and blends well with the landscape.

Fig. 1-46. Exposed finishes are also hard to clean because the texture is so rough.

Fig. 1-47. Exposed aggregate finishes, using ¾-inch gravel on the left and pea gravel on the right.

Fig. 1-48. A concrete job with the newer type etching design.

Fig. 1-49. Difference between a plain slab and an etched one. The etched design still requires a broom.

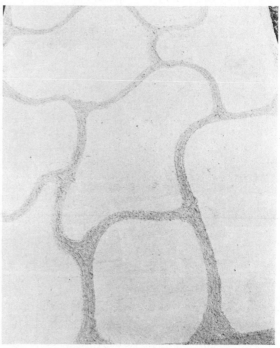

Fig. 1-50. Etching lines need not go very deep—only enough to create the design.

28

Fig. 1-51. All edges, expansion joints, seams, and stringers must be etched. All corners must be rounded.

Fig. 1-52. The etching lines will stay darker than the rest of the concrete, making the concrete look more like flagstones set in mortar.

Fig. 1-53. Brick insert to break up solid concrete look.

Fig. 1-54. Brick bordered walkway extends width.

Fig. 1-55. Patio bordered and sectioned with brick inserts for a warmer appearance.

Fig. 1-56. Brick inserts in the middle of the patio for design and possible borders for wooden planter boxes.

Fig. 1-57. Concrete must be set up and hard before the brick can be set.

Fig. 1-58. Unusual seam designs add a touch of flair to a slab.

Fig. 1-59. Patio slab equally sectioned by seams.

Fig. 1-60. For long seams, use a tool that will accept extensions.

Fig. 1-61. Planter in the middle of a patio slab.

Fig. 1-62. Mowing strip keeps grass from growing into the driveway.

Fig. 1-63. Mowing strip bordering a tree.

Fig. 1-65. A 2-inch-high piece of benderboard as an expansion joint.

Fig. 1-64. Circular mowing strip outlining a circular driveway.

Fig. 1-66. Walkway cracked where a seam or expansion joint should have been installed.

Fig. 1-67. Walkway tiles used to border a small patio slab.

Fig. 1-68. Using the edger with the arc side on the inside of the concrete will scribe a line that the seamer can follow. This will make an even border around the slab.

Fig. 1-69. Concrete ramp on the edge of asphalt at a driveway. The concrete will prevent the asphalt from breaking away.

Chapter 2

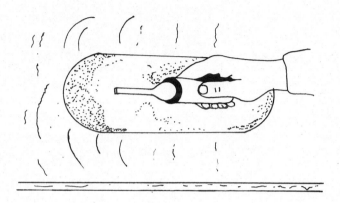

Concrete Costs

AFTER YOU HAVE SPENT SOME TIME DECIDING on the type of job you want, you'll probably want to get an idea of what it is going to cost. In this chapter I will explain how the concrete company bases its charges and what you will be expected to pay.

BASIC COST

Concrete is sold by the cubic yard, broken down into ¼-yard increments. One yard of concrete will cost anywhere from $35.00 to $50.00, depending where you live. If you were to order 1¼ yards, the basic cost would be (using the $50.00 figure) $62.50—$50.00 for the full yard plus $12.50 for the quarter yard. The concrete dispatcher will give you the appropriate figures on request.

On top of this basic cost, there will be added charges where applicable. Short load and overtime fees are normal, although the dollar amounts and times vary from company to company. After these, of course, comes sales tax.

SHORT LOAD

Short load fees are common, but the dollar amount varies. Generally, it will amount to $7.00 per yard for every yard ordered under seven. Most concrete trucks have a capacity of 7 to 8 yards, with some special rigs holding up to 10. For example, if you ordered 3 yards of concrete, the company will charge you an extra fee for tying up a truck on a small job when it could be delivering a full 8 yards. To compensate for it being unavailable, a short load charge is applied. In this case it would be $28.00, $7.00 times the other four yards it could have carried.

This cost is not hidden, and the concrete dispatcher will advise you of that when you contact him. All concrete companies have this charge, so don't feel that it is only aimed at the homeowner. Even professionals have to pay it (Fig. 2-1).

Short load charges are not applied to trucks carrying 7 yards. For the most part, many companies reduce this fee to 5 yards. Therefore, it

over 7 yards	no charge
7 yards	no charge
6 yards	$7.00
5 yards	$14.00
4 yards	$21.00
3 yards	$28.00
2 yards	$35.00
1 yard	$42.00

Additional delivery charges can be added for deliveries under 3 yards

Fig. 2-1. A sample list of short load charges.

would be in your best interest to talk to the concrete dispatchers from a few different companies to see who has the best deal.

OVERTIME

Concrete companies give customers a specific amount of time to unload the concrete. After that time, they can charge overtime. The allotted time is normally more than adequate to empty the truck. Again, companies vary in this, too. Some allow only 4 minutes per yard, while others might go as high as 10. While checking for short load charges, see how long you'll have to unload the "mud."

Overtime charges are based on minutes. For example, you might have to pay an extra 50¢ for every minute you have gone over the allotted time. This fee is to ensure that the concrete finishers are working hard. Time is money, and the more time that truck stays on your job, the less time it will have to make more deliveries. Even though 50¢ a minute seems like a lot, don't be discouraged and rush your job. An extra 10 minutes will only cost $5.00, and that might be all the time you need to catch up on screeding and have the job turn out right.

Many times the concrete truck driver is in control of overtime fees. Get on his best side. When he arrives, offer him a cup of coffee or a cold drink. Explain to him your plans and the general job layout. If he sees that you are on your toes and ready to go, he might overlook a few minutes of overtime. Your job must be properly set up, helpers present and ready, and all equipment available. Getting the

concrete out of the truck and into the forms will be the single most demanding and important phase of the job.

If you experience problems or the type of job you're doing is taking a little longer than expected, don't fret. The overtime bill won't amount to that much, and it is much more important to get the concrete placed correctly the first time. It is much better to pay a few extra dollars in overtime than to have to hurry and not get the concrete poured and screeded right the first time (Fig. 2-2).

CLEANUP

Short load charges are made on orders of less than 7 yards. Even if you order 10 yards and it takes two trucks to bring the concrete, you will not have to pay a short load fee. The only exception for additional fees on second trucks is on cleanups.

A cleanup load is one that is made after the first truck has made its delivery, and there was not enough concrete to do the job. This only applies to jobs where the original truck had the capacity to carry the full amount, but the customer made an error in his calculations and didn't order enough concrete. For example, if you ordered 3 yards of concrete and that was not enough to do the job, the truck coming with the cleanup load is going to cost you a pretty penny. Just for 1 extra yard of concrete, you might have to pay $100.00 (Fig. 2-3).

The logic for this is simple. If that one truck could have carried 4 yards, it should have been delivered at the original time. Because the truck has to make an extra trip back to the plant for just 1

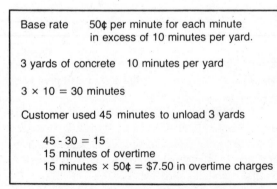

Base rate   50¢ per minute for each minute
             in excess of 10 minutes per yard.

3 yards of concrete   10 minutes per yard

3 × 10 = 30 minutes

Customer used 45 minutes to unload 3 yards

    45 - 30 = 15
    15 minutes of overtime
    15 minutes × 50¢ = $7.50 in overtime charges

Fig. 2-2. Example for calculating overtime charges.

```
3 yards of concrete ordered
Concrete selling for $50.00 per yard
3 × $50.00 = $150.00

Customer orders 3 yards, takes 45 minutes to
unload, and requests a 1-yard cleanup.

Basic charge:          $150.00 for 3 yards
Overtime (15 minutes):     7.50
Cleanup (1 yard):        $100.00
                         $257.50

Charges with no cleanup and correct order
Basic charge           $200.00 for 4 yards

Overtime (15 minutes):      7.50
                         $207.50

A savings of $50.00 with correct ordering
```

Fig. 2-3. Calculations for a job needing cleanup, including overtime.

yard, it is going to cost the company fuel, driver time, truck time, and delays in other scheduled deliveries. Be accurate when measuring square feet, be certain of the depth, and ask the dispatcher to calculate your figures for you.

On special jobs with angles, curves, or steps, order an extra ¼-yard of concrete. The cost is minimal, especially compared to a $100.00 cleanup.

Many smaller and some large concrete companies are very anxious to help the homeowner and infrequent concrete pourer. They will be glad to answer any questions you may have. I even know some concrete dispatchers that will come out to your job site, measure it, and calculate the yardage for you. This extra service is not always available, but if you want the help, all you have to do is ask. The worst they can do is say no (Fig. 2-4).

DRIVERS

Concrete truck drivers are regular people. They have been around concrete work for quite a while. Although most of them are not concrete finishers, they may be able to point out a few of your minor errors if there are any.

While you are offering the driver a cup of coffee and showing him the job layout, ask his opinion as to the forming and stake support. Having seen many jobs, he may have a hint or two on how your job may go easier and smoother. To go along with this courtesy, he may recognize your sincerity about doing a good job and really try to help. I have had drivers overlook the extra overtime charges and even help place the concrete on special occasions. Don't expect the extra hand, but be aware that courtesy can only help.

ONE-YARD HAULERS

If you are going to pour a very small job of 1 yard or less, there is a way to avoid extra short load charges. Many concrete outlets, both private and commercial, have small 1-yard concrete trailers available for rent (Fig. 2-5).

These heavy-duty trailers are capable of hauling up to 1 yard of concrete a limited distance from

Fig. 2-4. Concrete dispatcher calculating yardage with your measurements.

39

Fig. 2-5. A 1-yard concrete trailer. A hydraulic ram and chain are in front. Tilt the trailer toward the rear for concrete dumping.

the batch plant. They are rented out, but the only fee you pay is for the concrete. Most companies give you up to 2 hours to return the trailer, with no overtime charges.

You can order the concrete in ¼-yard increments at a set price. No other fees are expected. There are a few points that should be considered. First, the vehicle used to tow the trailer must have a heavy-duty trailer hitch. The temporary snap on hitches will not work. Concrete is very heavy and will damage the hitch, possibly losing control of the trailer and causing an accident, to say nothing of 1 yard of concrete in the roadway. Heavy-duty dock bumpers on pickup trucks seem to do a good job. I have hauled many a yard of concrete in those trailers towed behind my pickup.

This is the most economical way to purchase fresh concrete in ¼-yard increments, but there are drawbacks. Travel distance from the batch plant is generally limited to just 5 miles. As you travel, the trailer vibrates, causing the rock to sink to the bottom of the load. By the time you reach the job site, the material on the bottom will be very rocky, while on the top there will be soup. A soupy mix is one with a lot of water and cream and no aggregate. This condition is undesirable because the opening for the concrete is at the bottom or the top, and there is no way to remix the concrete.

If travel distance is less than 5 miles, you should have no problems. Another way to avoid aggregate sinking is to use pea gravel instead of the regular ¾-inch rock. The smaller size of the aggregate does not displace as much as the larger, and the load will be delivered in a much better condition. To go along with that, also request a five-to-six bag mix.

The bag mix refers to the amount of cement powder added to the concrete. Ordinary concrete comes with five bags of cement added to 1 yard of concrete. Some independent 1-yard hauler facilities only prepare a four-sack mix to their ready-mix concrete. Although that mix might be well-suited for some jobs, it has been my experience that a five- or six-sack mix is much stronger and helps to hold the concrete in a better condition during the trip from the plant to the forms.

These trailers provide a very good service and can help you save money. I have used them on many occasions and have been pleased every time. To locate the 1-yard hauler facility nearest you, check the Yellow Pages in your telephone directory.

ONE-YARD TRUCKS

To complement the concrete trailers, some outlets offer a 1-yard truck. Essentially, these rigs are miniconcrete trucks. A small drum is attached to the frame of a heavy-duty dual wheel truck and constantly rotates through a chain-driven hydraulic assembly. The beauty with this unit is that travel distance is almost unlimited. With the concrete constantly mixed, there are no worries about an uneven mix.

The cost for these trucks are a little higher, but well worth the price when travel is extended over 5 miles. I strongly recommend the use of either of these units when pouring 1 yard or less.

Chapter 3

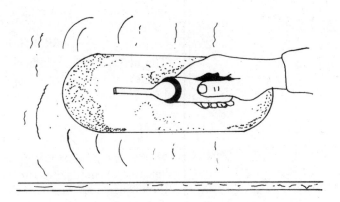

Calculating Yardage

THE MATHEMATICS USED TO DETERMINE THE amount of concrete you'll need are not difficult. Although concrete is sold by the cubic yard, square footage calculations are the determining factor.

CUBIC YARD

One cubic yard can be described as a cube measuring 3 feet by 3 feet by 3 feet (Fig. 3-1). To convert the yardage into feet, multiply the sides by each other in feet. For example, side 1 times side 2 (3 feet × 3 feet) equals 9 square feet. Take this product and multiply it by side 3 (3 feet × 9 square feet) equals 27 cubic feet (Fig. 3-2).

One cubic yard (a yard) equals 27 cubic feet. One yard of concrete will cover 27 square feet at a depth of one foot (Fig. 3-3).

Because your slab will be only 4 inches thick as opposed to one foot, you will have to divide those 27 cubes into smaller ones that measure 4 inches. To do that, multiply the 27 cubes by 3. Since 4 inches is one-third of a foot (12 inches), those 27 cubes will

cover three times the area when spread out at 4 inches in depth (Fig. 3-4). Therefore, that cubic yard will cover 27 cubic feet times 3, which equals 81 square feet (27 × 3 = 81). One yard will cover 81 square feet at a depth of 4 inches.

There are no variations from this basic formula. Don't try to squeeze any more concrete out of that cubic yard. If your slab measures 85 square feet and the depth is exactly 4 inches, one yard will not quite cover it (Fig. 3-5).

ONE-FOURTH-YARD INCREMENTS

Because concrete is sold by the cubic yard, concrete companies sell it broken down into ¼-yard increments. You can order any amount, but it will have to be totaled in one-fourth sections. For example, if you needed extra concrete to cover the 85 square-foot slab I mentioned before, you would have to order 1¼ yards of concrete. There will be some concrete left over that might be perfect for a mowing strip or trash can stoop that should be formed and ready to pour. The same would hold

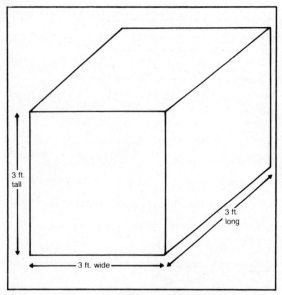

Fig. 3-1. A cubic yard measuring 3 feet by 3 feet by 3 feet.

true if you only had a slab that measured 70 square feet. You would have to order a full yard. Keep these figures in mind when designing the dimensions for your job. Attempt to keep the square footage as close as possible to a square footage that will be close to ¼-yard figures.

SHORTCUT

The number 81 is a difficult number to use as a divider. It will not fit evenly into even numbers, i.e., 200 divided by 81 will not produce an even answer. I have always had good luck using the figure of 80, rather than 81 (Fig. 3-6).

Using 80 as opposed to 81 has another benefit besides easy mathematics. It gives you a one-square-foot buffer per yard of concrete. By omitting the 1 square foot, you have gained a little concrete. This small amount could be a lifesaver on a job. If one of your helpers accidently dumped a wheelbar-

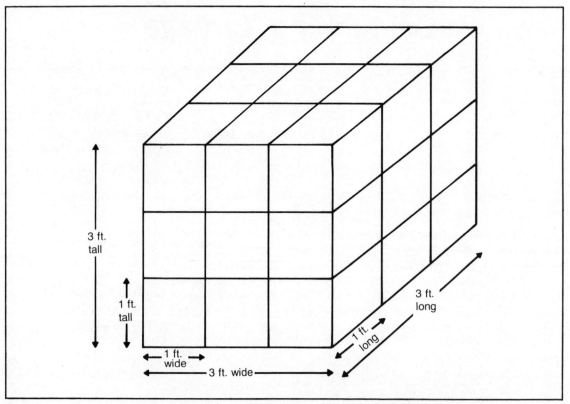

Fig. 3-2. A cubic yard evenly divided into 27 separate 1-cubic-foot blocks. Diagram is similar to a popular cube puzzle.

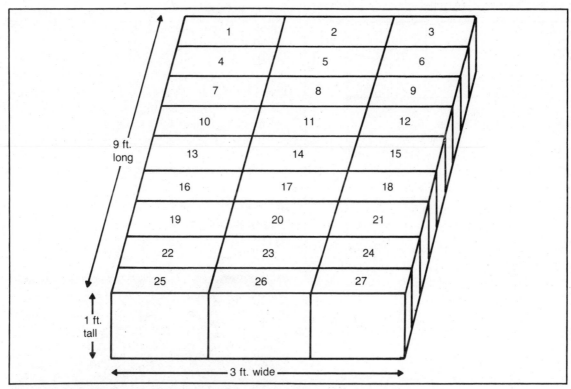

Fig. 3-3. Twenty-seven cubic feet laid out to cover 27 square feet at a depth of 1 foot.

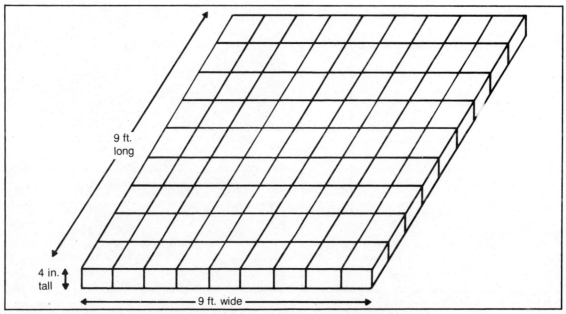

Fig. 3-4. One cubic yard of concrete will cover 81 square feet at an even depth of 4 inches.

43

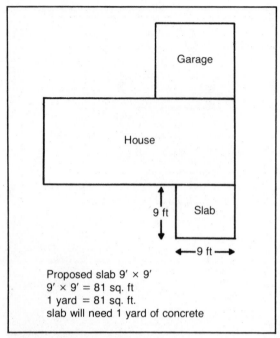

Proposed slab 9' × 9'
9' × 9' = 81 sq. ft
1 yard = 81 sq. ft.
slab will need 1 yard of concrete

Fig. 3-5. Sample plans of a concrete patio slab along with concrete yardage calculations.

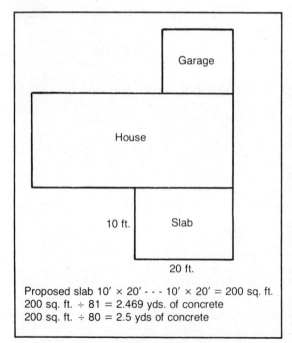

Proposed slab 10' × 20' - - - 10' × 20' = 200 sq. ft.
200 sq. ft. ÷ 81 = 2.469 yds. of concrete
200 sq. ft. ÷ 80 = 2.5 yds of concrete

Fig. 3-6. It is easier to divide by 80 rather than 81. The extra 1 square foot is a built-in buffer.

row on the way to the forms, the buffer might be enough to make up the difference.

When calculating yardage for your concrete job, divide the amount of square feet by the factor of 80. Use accurate measurements and always estimate over rather than under (Fig. 3-7). If one dimension of the slab measures 9 feet, 10 inches, use the figure of 10 feet. It is much better to have too much concrete than not enough (Fig. 3-8).

EXAMPLE

Let's say your slab measures 10 feet by 20 feet. To determine the amount of concrete needed, multiply the length times the width, then divide that answer by the factor of 80: 10 feet × 20 feet = 200 square feet. That is the amount of square feet included in

Fig. 3-7. Make your measurements accurately. Use a helper to hold the other end of the tape measure if needed.

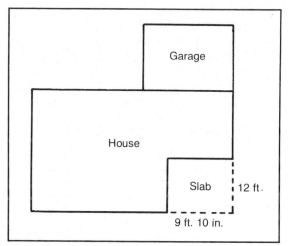

Fig. 3-8. Round off close measurements to feet. Twelve feet times 9 feet, 10 inches, is too hard. Twelve feet times 10 feet is easier and quicker and will not cause you to order any more concrete.

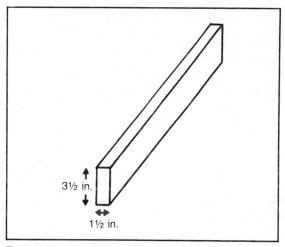

Fig. 3-9. Present day 2 × 4 lumber actually measures 1½ inches by 3½ inches. Lumber is sold in even lengths only, i.e., 6 feet, 8 feet, 18 feet, etc. Boards over 20 feet are sometimes a special order.

the slab. Because the area is 4 inches deep, divide by 80: 200 sq. ft. ÷ 80 = 2½, or 2½ yards of concrete.

Using this formula, you can be assured of enough concrete to complete your job. The depth of the job cannot be any portion over 4 inches. I cannot stress enough the importance of accurate depth. The majority of people I know who have come up short of concrete had accurate calculations. Their only error was in depth. Portions of their slab measured 5 to 6 inches deep in places, and that accounted for an additional amount of concrete needed.

The biggest concern of many do-it-yourself concrete finishers is ordering enough concrete. If you follow these directions and take precautions to maintain a 4-inch depth, you will not have any problems. To add to your reassurance in ordering enough mud, I give you another tip. Most 2 × 4 form lumber that is sold now actually measures 1½ inches × 3½ inches. The depth of a slab accurately graded will measure between 3½ inches and 4 inches. Don't be concerned about the strength of a 3½-inch slab. I have poured many at that depth, and they are all still in good shape. The extra ½ inch will, however, give you another buffer.

The actual dimensions of the form lumber

Fig. 3-10. Concrete color dust is bought in bulk by the concrete plant and sold to customers in 1-pound bags.

normally do not pose any special problems. The only exception will be in areas of particular custom forming. These would be jobs that include 2 × 4 stringers and special projects using the dimensions of lumber in specific plans. Those unique areas will be explained in a later chapter (Fig. 3-9).

COLOR

Colored concrete comes in two ways—from the plant already mixed or by the finisher adding the dust after the concrete is on the ground. Color charts and cost figures are available at the concrete batch plant, and the dispatcher will help you in your decision. Adding color is expensive. Depending on the dye, you can expect to pay an additional $8.00 to $50.00 per yard if the concrete is to be fully colored at the plant. Applying the dust yourself will save money mainly because you will use less.

Generally speaking, the color dye dust is sold in 1-pound bags, with 1 pound adequately covering 20 square feet (Fig. 3-10). The dust will be applied and floated in after the concrete has been tamped and floated. It is a time-consuming chore and very messy. The dyes are very potent and will stain anything with which it comes in contact. Caution must be used in its application.

If colored concrete is in your plans, contact your local concrete company and ask for advice, color charts, and prices. The dispatcher can also suggest the best way to seal your colored slab and give you pointers on pouring for your climate region.

Chapter 4

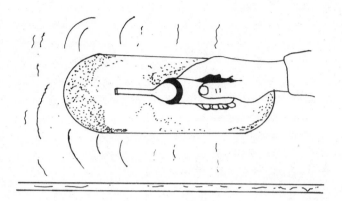

Ordering Concrete Delivery

PLACING AN ORDER FOR CONCRETE REQUIRES you to furnish the dispatcher with certain information. He will need the yardage and other information.

AGGREGATE

The rock used in mixing concrete comes in various sizes. Normal pours where the concrete is directly dumped from the truck to the forms contains ¾-inch gravel, unless the order specified different. You can get pea gravel at no extra charge if you need that size for a special reason (Fig. 4-1).

The aggregate is added to the mix as determined by the batchman who regulates the conveyor belts that feed the hopper (Fig. 4-2). There are no special procedures involved to provide the aggregate you designate.

There are certain times when you will need to specify a load of concrete to be other than the norm, such as when you use a concrete pump. I will discuss pumping procedures later.

MIX

Concrete is sold by the cubic yard and is broken down into ¼-yard increments. It is mixed by weight. The batch operator uses a predetermined chart to gauge the amounts of materials added to mix a concrete load (Fig. 4-3). In hundreds of pounds, sand and aggregate are added to gallons of water and a prescribed amount of cement.

Ordinary ¾-inch aggregate is mixed with sand and five sacks of cement. Each sack contains 90 to 94 pounds of powdered cement. Five gallons of water is added to this mix. Conveyor belts deliver the sand and gravel to a hopper. On top of the hopper is a large storage bin filled with powdered cement (Fig. 4-4). At the base of the hopper, all of the ingredients are mixed and are dumped with water into the concrete truck. The rotating action of the drum on the truck mixes the batch and produces fresh concrete (Fig. 4-5).

A five-sack-per-yard mix is more than adequate for most jobs. Some areas require a stronger

Fig. 4-1. Three-fourth-inch aggregate is much larger than pea gravel.

Fig. 4-2. Conveyor belts feed sand and gravel to the mixing hopper from storage piles.

Fig. 4-3. Concrete dispatcher operating controls to load truck.

mix of cement because of extremely cold winter temperature. Your concrete dispatcher will inform you of any standards when you contact him.

A six-sack mix produces very strong concrete and seems to be easier to finish because of the rich mixture of cream. Unless specified, a six-sack mix is not necessary for a ¾-inch gravel load. If you order pea gravel, though, get the extra sack per yard.

Pea gravel can be used to produce a fine looking exposed aggregate slab. It is also an essential aggregate for use in concrete pumps. If you hire a concrete pump, you must order a pea gravel six-sack mix. The extra sack of cement per yard helps to add strength to the smaller rock and also helps the concrete to flow through the small diameter hose.

Fig. 4-4. Mixing hopper with cement storage bin above.

An extra sack of cement per yard will also cost you more money—generally from $2.50 to $5.00 per yard. Concrete costs are calculated on the amount of material used in the mix. So, understand-

ably, the concrete company must cover its cost.

PLANNING THE POUR

The type of pour you plan will determine the mix

Fig. 4-5. Concrete being loaded into concrete truck.

ordered and the amount of help you'll need. Helpers are an integral part of the job. Professionals can handle a small pour by themselves, but a novice concrete finisher should have help no matter the size of the job.

If you are going to wheelbarrow 2 to 3 yards of concrete from the front to the back of your house. I suggest having at least four and preferably five workers. You'll need two men to run the wheelbarrows, two men screeding, and one to help move the concrete around and assist as needed. If you're wheeling any more than 4 yards, you should have another man and another wheelbarrow.

Normal wheeling requires a regular three-fourth five-sack mix. If you want anything other than that, it will be because of a design characteristic or because of weather factors in your area.

The other time when a normal mix is adequate is when you pour directly from the truck to the forms. You won't need as many helpers, but I again

Fig. 4-6. A common 7- to 8-yard concrete truck.

suggest at least four workers: one to work the chute from the truck, two screeding, and one to move concrete inside the forms for the screed men.

TRUCK MANEUVERABILITY

When pouring directly you must also be concerned about the ability of the truck to maneuver around the job site. Most common are the 7- to 8-yard rigs (Fig. 4-6). Drivers of these vehicles can get into and out of some of the tightest places. Don't make it too hard on them. If you are in doubt as to the ability of a rig to fit where you'll need it, give the concrete dispatcher a call. He will give you the dimensions needed for the truck to maneuver.

Fig. 4-7. An extra large 10-yard capacity concrete truck, with extra wheels down.

Larger trucks are also available in many areas. These rigs can haul up to 10 yards of concrete at one time (Fig. 4-7). The added weight of 3 yards of concrete requires these vehicles to have an extra set of wheels (Fig. 4-8). These wheels are hydraulically raised and lowered. They are placed on the ground while the truck is on the road with more than 7 yards. When it arrives at the job site, the wheels are raised out of the way. Even with the wheels raised, this particular truck is very long. Make sure there will be enough room for this rig to operate if it is sent to your job.

Ten-yard concrete trucks normally are sent only to jobs where more than 8 yards of concrete are ordered. Not all companies have them. While talking with the concrete dispatcher, let him know what kind of a job you are planning and how the concrete will be placed in the forms. Let him know of any tight maneuvering spaces and advise him if the 10-yard truck will not work.

PUMPING

On jobs requiring 5 yards or more of concrete that has to be wheeled, I have always used a concrete pump. Independent pump owners and some concrete companies operate special pumps that get the concrete from the truck to your forms quickly and easily. I think these pumps are fantastic.

Their original purpose was to grout retaining walls. A smart block mason used the pump to place the concrete for a slab. Since then, many concrete finishers have used pumpers successfully.

The pump owner/operator usually will man the end of the hose and place the concrete. You will only need two men to screed and another to pull hose (Fig. 4-9). Jobs seem to go much faster, cleaner, and easier when done with a pump. The only drawback might be in the cost. Concrete pumps usually have a base fee of between $50.00 to $65.00. Also, they may charge $5.00 per yard for every yard pumped over seven. The extra $5.00 per yard might sound high, but time is money. The longer a pump sits at your job, the less amount of time it will have to work at another job.

Another cost increase will be in the mix. As mentioned before, common concrete pumps require the use of pea gravel as opposed to ¾-inch rock. With the pea gravel, I strongly recommend using a six-sack mix. This will increase the concrete fees by $2.50 to $5.00 per yard. For a 7-yard job that is pumped, you can expect to pay an additional $65.00 to $100.00, depending on the actual cost of the pump operation and the going price for cement.

Even with the added cost, you will still be saving money by doing it yourself. If you were going to pump the job, a professional probably would have, too. You won't be saving any money with the professional, and the added cost of a pump job will be appreciated when the work is done.

DATE TO POUR

Deciding on what day to lay your slab depends on the time all forming will be done, the availability of helpers, and the schedule at the concrete company. You can usually determine when the forming will be completed. Your helpers may not be available on certain days, so you will have to confirm plans with them.

Concrete companies regularly schedule jobs as they come in. They have certain contractors who have standard pouring times every day. Around them, they usually have ample space to deliver your concrete (Fig. 4-10). Check with the dispatcher at least two days ahead of time. It's not a bad idea to check a week ahead of time. Confirming a date for delivery will ensure your concrete delivery on the day and time you requested. The only delays will be caused by mechanical breakdown or an added amount of time taken to unload the delivery ahead of yours.

For the most part, weekday deliveries are the best. That is when the plant is operating with a full complement of manpower and the most trucks are on the road. Most companies make deliveries on Saturday, but they must add an extra fee. The fee doesn't amount to very much, generally from $5.00 per yard, or a flat fee of around $15.00 to $20.00. This charge is not standard and can vary from company to company. The extra money helps to pay for any overtime wages to employees working more hours than their standard workweek.

Fig. 4-8. Close-up view of extra wheels on a 10-yard concrete truck. Wheels and chutes are hydraulically raised when it is time to unload.

Fig. 4-9. Pumping a concrete job. Pump operator manning the hose while helpers pull back the hose and screed concrete.

TIME TO POUR

The time to pour is not critical, but there are some factors involved. During long hot summer days, it is best to lay the slab as early as possible—7:00 A.M. to 9:00 A.M. Because the afternoons get so very warm, the concrete can set up too fast. When that happens, you stand a chance of losing the slab.

The concrete will lose its moisture too fast, and you will not have ample time to effect the finish. By pouring early in the morning when the temperature is cool, the concrete will have more time to set up. You will have ample time to finish.

During the winter months, just the opposite is true. Again, you should schedule the pour early in the morning. Although you will have little worry of the concrete setting up too fast, it might take all day for it to get hard.

Cooler and damp weather won't evaporate the moisture in the mix very fast. It may take several hours for the surface to set up enough to finish. Many concrete companies mix winter concrete using hot water. This helps the evaporation process and cures the mud sooner.

Planning the job for an early pour is best. I have delivered concrete at 4:00 P.M. on a summer's afternoon and had to put on the final finish at 8:00 that night under lights. Finishing concrete under artificial light is no fun.

WEATHER FACTORS

You cannot finish concrete in the rain or snow. Rain drops and snowflakes disturb the surface, add water to the cream, and make finishing impossible. If the threat of rain seems possible, cancel delivery and reschedule. Concrete dispatchers don't like to cancel, but they also don't have to finish the slab.

I have rescheduled certain jobs as many as five times due to rainy weather. That's because rain will

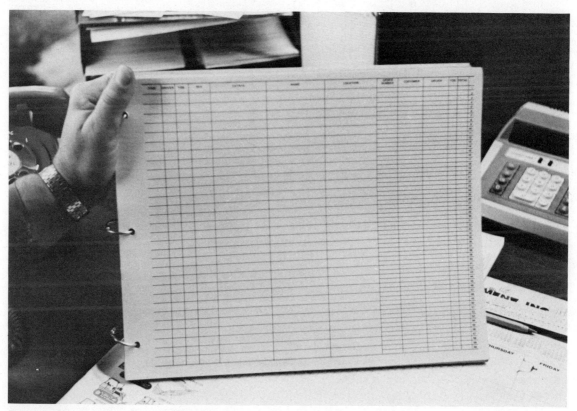

Fig. 4-10. Concrete dispatcher's scheduling book.

Fig. 4-11. Concrete truck driver in radio contact with concrete dispatcher's office.

ruin a slab. Don't feel bad if you cancel and it doesn't rain. It is much better to be safe than sorry. Once concrete is on the ground, it's there to stay.

Never be afraid to cancel an order. Dispatchers can reschedule you at a different time. Do not wait until the last minute to call, though. Sometimes, the concrete trucks are running ahead of schedule. On a rain-threatened day, there might be other cancellations, and your load might be on the way an hour early. Call the dispatcher as soon as you decide to cancel. Once the concrete is loaded

into the truck, there is no turning back (Fig. 4-11).

ORDERING

Tell the dispatcher that you want to place an order for concrete. He will ask your name, address of the job, and your telephone number. Then he will want to know what kind of mix you'll need. You will have to know whether to order a ¾ five-sack or a pea gravel six-sack mix. He'll need the amount of concrete you want. If you are still not quite sure as to how many yards are required, give him the dimen-

S·O·C
№ B 6296

Southern Oregon Concrete, Ltd.

PHONE (503) 476-1471 ● 689 UNION AVE. ● GRANTS PASS, OREGON 97526

Customer's
Order No._____ Date_____ 19____

Sold to:_____

Address:_____

Deliver to:_____

DRIVER	EQUIP	ORIGIN	PLANT SLUMP	JOB SLUMP	WATER REQUESTED

ON JOB		START POUR		END POUR		TEMP		C.O.D. ☐ CHARGE ☐

CONCRETE IS CAUSTIC AVOID CONTACT WITH SKIN

CUBIC YARDS	CONCRETE	
ACCELERATOR		
HOT WATER		
SAND GRAVEL		
CRUSHED ROCK		
DELIVERY CHARGE/MILEAGE		

A cash discount of_____may be deducted if remittance is made on or be-fore_____. Discount will be allowed only if all previous charges have been paid.	**TOTAL**

NOTICE: A lein may be claimed for all materials/supplies delivered to the property. Delivery policy F.O.B. jobsite, curbside only. If delivery inside curb is requested the undersigned will not hold S.O.C., Ltd. responsible for any charges or liabilities incurred beyond curb side; the undersigned will bear responsibility, i.e. wreckers, fees, property damage.

_____
Accepted for Purchaser

Concrete unloading time in excess of 10 min. per/yard will be charged at the rate of 75¢ per minute or any fractional part thereof.

Service charge of 2% per month (24% per annum) will be added if not paid by the 25th of the month following date of invoice.

Fig. 4-12. Concrete order form. You will be asked to sign this when the truck is ready to leave.

sions of your slab and let him calculate it. If your figures are the same, you're in business.

After that information, you will be asked on which day you plan to pour and at what time. If that fits into his schedule, you are in. If not, you'll have to replan the job. This is why it is a very good idea to order a week ahead of time to ensure delivery on the day and time you prefer.

If you want color added to the entire mix, you had better notify the dispatcher at this time. A plant may be out of a certain color dye. An order then must be placed for a new supply. Depending on the company and its location, this could take as long as a week.

While you are talking to the concrete dispatcher, he is making notes and filling out a schedule slip and order form (Fig. 4-12). After you have discussed the order, go over the date, time, address, phone number, and yardage again.

Remember that the concrete dispatcher is responsible for the scheduling of the trucks and the proper mixing of the concrete. He wants you to be satisfied with his delivery just as much as he wants your business. Courtesy shown toward him will

Fig. 4-13. Making a predelivery call to your concrete dispatcher is a very good idea. It can help both of you.

reward you with plenty of information. The concrete business relies on return customers. If you are happy with the company's service, chances are that you'll order from it again. The concrete company also hopes that you will tell your friends and neighbors about its quality service, too.

PREDELIVERY CALL

On the morning of the pour, call the concrete company and once again confirm the order. This helps out both you and the dispatcher (Fig. 4-13). By confirming the order, you let the dispatcher know that your job site is ready for delivery and you are prepared for the load. This confirms his schedule and gives him an added element of on-time service.

For you, this call makes sure that the address, mix, yardage, and time are what you ordered. I have done this on all my jobs. Calling an hour or two before delivery gives me an idea on how his schedule is going, too. At times there will be a cancellation, and I could get the concrete earlier. When the job is ready to go, an early pour is really nice. The job is done sooner and gives me more time to sit back and admire my work.

Chapter 5

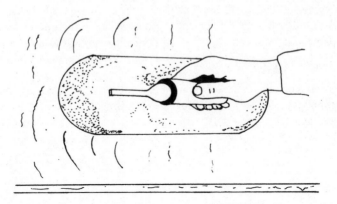

Forming and
Grading Tools and Materials

LAYING A CONCRETE SLAB OR WALKWAY CON-
sists of two basic phases; forming and grading
and pouring and finishing. In this chapter I will
discuss the different tools needed to do the forming
and the material needed for its effect (Fig. 5-1).

GRADING

Grading refers to the leveling of the ground (base)
for the slab. A flat base will offer the greatest
support for a slab and will help to prevent cracks.
An even base also ensures correct calculations
when determining concrete needed.

Heavy-duty shovels are a must. Both the round
point and square point have their separate uses. A
square point is used more than the round. The
straightedge is useful in squaring off sides and
scraping the base to grade. A rake is quite useful for
moving dirt, filling in, and removing small amounts
of dirt from the surface. The rake is also handy for
moving wet concrete. A good pick is essential if you
have to remove more than a couple of inches of
grade. The type of pick I refer is one with a point on

one end and a blade on the other. A hoe can be
useful, but it is not a must. I wouldn't suggest
buying one just to help with one concrete job.

One of the most important tools in concrete
work is a heavy-duty wheelbarrow. A contractor's
type with an air-filled tire is the best. A garden
wheelbarrow with a solid wheel will not stand up
well. If you don't already have such a wheelbarrow,
one can be purchased at hardware stores or lumberyards. You can rent one from a local rent-all
facility. Wheelbarrow rentals are not expensive. If
you don't really need one for grading, though, just
rent it for the day you pour.

Grading can be a backbreaking chore, or it can
go rather smoothly. Generally, I form the job first
and grade last. This procedure gives me a definite
line to grade to. There is not much sense in trying to
grade when you're not even sure what the grade will
have to be.

FORMING TOOLS

Forming consists mainly of carpentry work; saw-

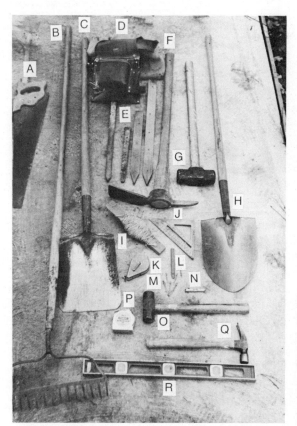

Fig. 5-1. All the tools needed for forming and grading: (A) handsaw, (B) rake, (C) square-point shovel, (D) nail bag, (e) wood and/or metal stakes, (F) pick, (G) sledge hammer, (H) round-point shovel, (I) string, (J) square, (K) chalk line, (L) carpenter's pencil, (M) double-headed duplex nails, (N) string level, (O) 3-pound stake hammer, (P) twenty-five-foot tape measure, (Q) claw hammer, (R) level.

ing, nailing, and supporting boards. Claw and sledgehammers, a tape measure, level, handsaw, square, pencil, chalk line, string, a nail bag, and Band-Aids (if you are prone to hitting your fingers with the hammer).

Claw and Sledgehammers

The claw hammer will be used to pound nails, and the 3-pound minisledge will be used to drive stakes (Fig. 5-2). A good claw hammer is capable of driving stakes, but the three-pound sledge does the job faster and with more accuracy. Also, the broader head on the sledge won't ruin as many wood stakes as the smaller one on the claw hammer.

Tape Measure

A heavy-duty tape measure makes the job easier to do. A 25-foot tape is almost perfect for most jobs. The wide-styled blade will stay in place and will not bow up to 7 feet. It is easier to use when trying to get an accurate measurement of a long side.

Fifty-foot tape measures are very useful when planning a walkway job over 25 feet. These tapes can be expensive. If you don't have one for your long walkway job, a 25-foot tape will work fine.

Saws

An ordinary handsaw is adequate for the amount of form cutting you'll have to do. On the other hand, if you are planning to use a lot of 2 × 4 stringers, a power circular saw might be easier. Some smaller

Fig. 5-2. Claw hammer and 3-pound minisledge hammer.

jobs won't even require the use of a saw. You can let certain forms go past the mark and extend out past another form.

Square and String

The square is a vital tool. Without it no job would ever match squarely with the house or anything else. This tool will be used to line up outside forms with the house, align corners, and place stringers. There are many types of squares available, and most any one will do. I suggest using the large type—the kind that measures about 1 foot by 1½ feet.

Heavy-duty string also aids in the squaring process. After a couple of guide stakes have been placed, a tight string between them will act as a guide for the form and other stakes.

Level

The string level is an inexpensive and handy tool and is used with a tightened string (Fig. 5-3). A taut string with this level attached can be moved up or down at either attached ends. The level will signify when the string is perfectly level. This guide helps to determine slope on long forms and also helps when attempting to establish grade. The cost is minimal for this tool.

Regular 2-foot-long levels are other important tools. Without a sturdy level, there would be no way to determine a proper grade. The small torpedo-type levels might be fine for very small projects around the house, but for a large concrete job you'll need a bona fide level.

Chalk Line

A chalk line is a roll of heavy-duty string inside a metal case. Colored chalk is added into the case and adheres to the string. When the string is pulled from the case, stretched along a solid surface, and then snapped, it leaves a straight line on the surface that is the color of the chalk.

A chalk line is not always necessary. When pouring long walkways, though, it really comes in handy. Lines can be chalked to show grade, point out the line where the top of the concrete will go, and be used as guides for screeding. Chalk lines during the forming phase are very useful in establishing high points for forms. These tools are also rather inexpensive. They can be purchased at local hardware stores and lumberyards. Besides their usefulness during the concrete job, they are also ideal for some projects around the house (Fig. 5-4).

Fig. 5-3. String level attached to a tight string.

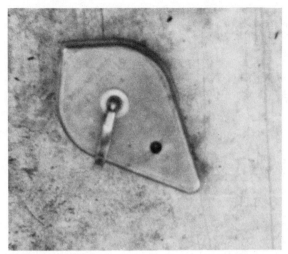

Fig. 5-4. Chalk line.

Pencil

A heavy-duty carpenter's pencil is very rugged and holds up well in construction. The wide lead protects it from breaking off. Pencil marks will be made on all kinds of things during the forming phase—from marking forms to making marks to establishing grade on the side of the house. Ordinary pencils will do the job, but when buying your form lumber from the lumberyard, ask about the pencils. The lumberyard person may give you one.

Nail Bag

A nail bag is very handy when forming. This item is a luxury and is quite expensive, so don't go out and buy one just for a single concrete job. If you do get one however, be sure to buy the type with a "goodie" pouch and one or two plain pouches. The leather bag with all the compartments is good for holding your tape, string level, pencils, razor knife (for sharpening the pencils), and chalk line. The plain bags hold nails and hammers. The plain ones can also carry the Band-Aids.

MATERIALS

Basic forming material consists of 2 × 4 form lumber and stakes. A good grade of Douglas fir wood is perfect for forming. The most important point to watch out for in choosing your forms is their straightness (Fig. 5-5). If a form slightly bows along the 4-inch side, it might be usable. Although not perfect, this form could be straightened out during the staking. If, however, the board bows across the top on the 2-inch side, discard it. It will not straighten out and is useless when forming concrete (Fig. 5-6).

Lumber comes in even lengths: 6 feet, 8 feet, 12 feet, 18 feet, etc. If the dimensions in your slab are an odd number, 9 feet for example, you will have to buy a 10-foot board. That extra foot might be wasted for the concrete job, but if you're like me, you will always find a use for it.

Boards over 20 feet are sometimes a special order item. The longer the board, the more likely it is to be warped. If you're going to form a job that's

Fig. 5-5. Checking 2 × 4 form board for bows. Stakes can straighten out some small bows.

Fig. 5-6. Looking down the top of a 2 × 4 for bows along that edge. Stakes cannot take out bows along the top edge.

24 feet long, I suggest using two 12-foot boards rather than one long 24-foot one.

Knots

Knots on the sides and edges of a board can mar a concrete finish (Fig. 5-7). Concrete will flow into them and harden. If you have the opportunity to select your forms, be sure to avoid lumber riddled with knots. If you have a form at the house that you plan on using, and it has a knot on one side, make sure that you form that knot on the outside of the slab. That way no concrete will touch it, and the job will turn out perfectly.

Benderboard Forms

If any part of your proposed job plan shows curves or rounded corners, you'll need to use benderboard forms. These thin strips of wood are approximately ⅜-inch thick and come in 4- to 6-inch widths. There lengths are generally from 6 feet through 12 feet. To form curved walkways or patios, use the 4-inch width.

For extremely tight curves, this brittle wood should be soaked in water. The added moisture will greatly add to the elasticity of the board and will enable you to form a very tight curve. Use caution with benderboard. The thin texture makes it very susceptible to breakage. When used as forms, they must be braced by 2 × 4s or dirt. See Chapter 11.

Forming slight curves on long walkways and driveways is difficult with benderboard. For these instances, I suggest using a 1-inch-wide board, and 1 × 4s are also available at the lumberyard. The sturdy quality of these larger boards makes forming easier. They are more rugged, so breakage is not as much of a concern.

Stakes

You should have plenty of stakes. Don't try to cut corners. The only thing worse than rain falling on your unfinished slab is to have a form or two break loose.

Stakes are sold at lumberyards and some nursery outlets. I recommend using 12-inch stakes in hard ground and up to 24-inch ones in soft ground. The stakes are holding your forms in place. Make sure they are long enough to support the forms and that they are pounded into the ground far enough to ensure against form movement once the concrete is placed.

I place my stakes every 3 to 4 feet. Never place support stakes further than 4 feet apart. Concrete is a very heavy material, and unsupported forms can be moved when concrete is poured against them. As a general rule, place your stakes at 3-foot intervals.

In areas where the ground is very hard or rocky, many professionals use steel stakes. The stakes are expensive, but they are worth their weight in gold to a concrete former who consistently finds himself working with exceptionally

Fig. 5-7. Checking for and discarding lumber with large knots on the tops and edges.

hard ground. With this type of base condition, a wood stake will split or break. It is almost a waste of time to try and form a job with wooden stakes if every time you go to sink one, it breaks.

If the ground in your yard is not rocky or terribly hard, don't go out and buy a few dozen steel stakes. If wood stakes will not work, check around and see if you can rent some for a few days.

If no rental yard carries steel stakes for your granite hard ground, you might be able to use lengths of rebar for stakes. *Rebar* is the round steel rods used to support block walls and chimneys. I'm sure you have seen some of it sticking up out of the ground at the base of a chimney or retainer wall under construction.

These steel rods can be cut to length and used as makeshift stakes. You won't be able to drive a nail through them because, unlike regular steel stakes, they won't have any holes drilled in them. You can drive a nail into the form next to them and then bend the nail around. These stakes are an alternative and do have their drawbacks. They do work in an extreme situation.

EXPANSION JOINTS

Expansion joints are complete separations in the concrete that allow for expansion and contraction in a slab. Concrete will expand during hot weather, resulting in cracks if allowance is not made for this expansion. These joints are not necessary on slabs that are open on two or three sides. Where a slab joins with another piece of concrete on more than one side, however, you should install an expansion joint. Walkways require an expansion joint every 10 feet or so. This is due to the long length and rather thin width.

Expansion joint felt is required whenever a

Fig. 5-8. Expansion joint felt and rebar.

private slab comes in contact with a city sidewalk. This material is composed of a feltlike board treated with a petroleum sealer (Fig. 5-8). The felt will work well, but it is difficult to work with. It must be placed inside the forms and braced with a 1 × 4 or 2 × 4 and stakes. The boards are pulled after the concrete has been poured around it. Edging can really be troublesome.

The sharp point of the edging tool sometimes digs into the felt and ruins a good edge, making a mess along with it. You might form the felt a little high. When the concrete is cured, simply use a razor knife to cut off the excess. Sitting high in the concrete will allow ease in using the edger. The tool will glide across the side of the felt and not dig into the top of it.

FELT ALTERNATIVES

If expansion joint felt is not required for your concrete job, you might consider using one of two alternatives. A piece of 2 × 4 redwood or pressure-treated fir will serve the purpose of an expansion joint. It will be easier to install and will be much easier to edge against. The board essentially will be a small stringer.

Redwood and pressure-treated fir will not rot. Once in the concrete, a piece of 2 × 4 will be almost impossible to take out and replace. Therefore, you will want to use something that will hold up over the years. The 2 × 4 can also be stained to match the surrounding wood on your house.

The simplest expansion joint material to use is benderboard. By splitting a 4-inch piece down the middle, you'll end up with a 2-inch-wide strip of wood ⅜ inches thick and cut to fit. After the concrete has been tamped and floated, simply wriggle the board into place. You may have to use a hammer to help pound it in, but it will fit. Afterward, you should use a hand float or trowel to smooth out the concrete disturbed by its placement.

The 2-inch piece of benderboard will not completely separate the concrete. It will guarantee that the concrete will crack underneath the board and not somewhere else.

Benderboard strips work fine on 3-foot-wide walkways. On widths any greater, the wood is just too thin to handle. It will be very difficult to insert a ⅜-inch piece of wood into a 4-foot-wide walk and keep it straight. Use a 2 × 4 in these cases.

NAILS AND NAILING

The most common and useful nails for concrete forming are *duplex nails*. These double-headed nails are easily pulled when it comes time to "strip" the forms. The first head on the nail shaft is butted to the surface of the wood to which it is nailed. The second head, located at the top of the nail, sticks out just enough for you to get the claw on a claw hammer around it. These nails come in different sizes: 6d, 8d, 10d, and sometimes larger. I have had good service out of the 8d or 8-penny nail. It is big enough to go through a wood stake and almost completely through the 2-inch side of a 2 × 4. Note that 6d nails are a little too small and do not hold the form as securely, and 10d nails are just too big for most forming jobs.

On ordinary forming jobs, the nails will be driven through the stake first and then into the form. This is because the stake will be on the outside of the form. Nailing from inside the form would make nail removal impossible after the concrete has been poured.

When using 2 × 4 forms, you will only nail from the inside when you are trying to secure a form located on top of a piece of existing concrete. For example, you want to pour a slab off the back of your house. A sidewalk runs along the side and will be next to the new slab, but you want the new slab to be 4 inches higher than the walkway. How will you place a 2 × 4 form on top of the existing walkway and secure it? Place stakes right next to the walkway—the exact place the side of the new slab will be. Place a 2 × 4 next to the stakes and on the outside (the stakes will actually be inside the forms). With the form in place, drive a 16d sinker nail through the stakes and through the forms. With all the nails driven, bend them over on the back side of the form (Fig. 5-9).

The bent nails will hold the form in place. After the concrete has cured, straighten the nails. Gently pull the form away from the stakes. The stakes will be embedded in concrete on three sides. The only

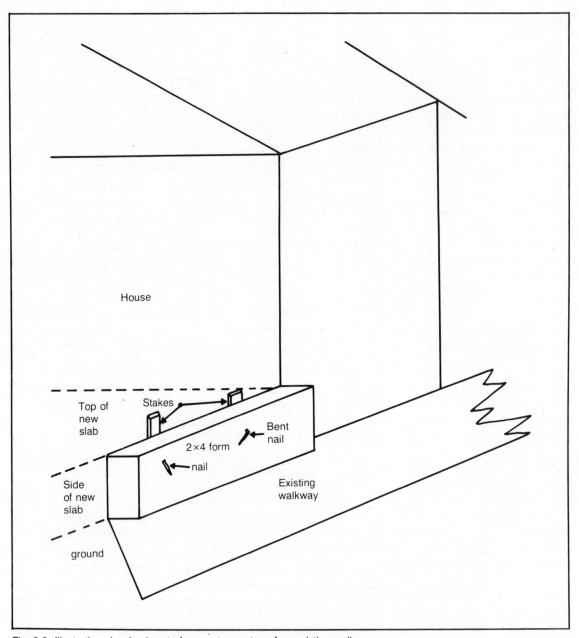

House

Top of
new
slab

Stakes

Bent
nail

2×4 form

nail

Side
of new
slab

Existing
walkway

ground

Fig. 5-9. Illustration showing how to form a step on top of an existing walkway.

side bare will be the one against the form. Pull the stakes out of the concrete and begin repairing the concrete.

Redwood benderboard is thin and brittle. Securing it to stakes it not an easy job. The best way I have found is to nail them from the inside of the forms with roofing nails.

A roofing nail has a very large head. The extra surface holds the benderboard in place, and not too many strokes of the hammer are needed to drive one in. Use ¾-inch roofing nails. They are long enough to secure the form, but not too long to make pulling the stake from the form difficult.

In the case of benderboard, stakes are placed on the outside of the form, and the nails are driven inside. Pull the stakes first before the nails. After the stakes have been removed, the nail points will be sticking out from the form. Use caution.

DRAINAGE PIPE

Plastic drainpipe is very good for directing water runoff. During the forming process, provisions should be made to insert this pipe where needed.

Flower beds between walkways and the house can get flooded during a heavy rain storm. To prevent the flooding, plastic PVC (polyvinyl chloride) pipe can be installed under the concrete, allowing an escape for the trapped water.

Drainpipe is also advisable under slabs that will block water runoff from the back of the yard to the front. For example, if you pour a storage shed slab extending from the side of the house to the fence, how will rainwater get to the street from the backyard? The water will tend to puddle next to the slab, eventually flooding the storage area. By simply placing a piece of drainpipe under the slab, the water will run to the front and out to the street.

Plastic drainpipe is relatively inexpensive and is available at hardware stores and lumberyards. When placing it under a slab, be sure it is buried under the concrete. Do not let it sit up into the concrete. That will cause a weak spot and possibly a crack. The best way to install it is to dig a trench into the base for it to fit.

Chapter 6

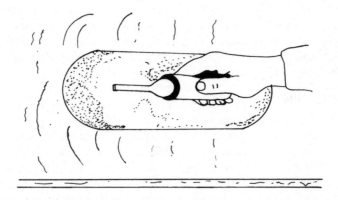

Pouring and Finishing
Tools and Materials

M OST OF THE SPECIALIZED EQUIPMENT CAN BE rented at local rental yards. The cost is minimal. If you had to rent every tool listed, I don't think the total bill would come to over $30.00 or $40.00. These tools can also be purchased at local hardware stores and lumberyards (Fig. 6-1).

RAKE AND SHOVEL

The rake and shovel are necessary tools. A heavy-duty rake can be used to push and pull concrete toward and away from the screed board. Concrete is heavy—too much for a flimsy rake.

A square-point shovel works well to place concrete. The flat nose guides concrete directly into the forms. A rounded shovel will allow too much concrete to fall over the sides, making for a messy job. A round point is perfect for keeping concrete inside the rounded chute from the truck.

WHEELBARROW

If you plan to wheelbarrow the concrete, use a sturdy, heavy-duty, contractor's wheelbarrow.

Never use a lightweight garden wheelbarrow with the solid rubber wheel. The little ones are fine for chores around the house, but they are unsuitable for working around concrete.

Using the wrong wheelbarrow will ruin it and also cause you to lose loads along the way. That will waste concrete and make for added standby time.

Contractor's wheelbarrows are reinforced for strength, have an air-filled tire, and can carry up to 5-¾ cubic feet of material. Don't however, try to handle a full wheelbarrow load of concrete immediately. Trying to control a large load like that without practice is asking for a fall. When you start out, tell the driver that you are not familiar with wheeling concrete. Have him fill the wheelbarrow only half full. This will give you a chance to understand what wheeling is all about. After you have gotten used to the load, weight, and balance, you can have the driver fill it a little more.

The air-inflated tire on a construction wheelbarrow is perhaps its best attribute. It will easily roll over small rocks and bumps and acts much like a

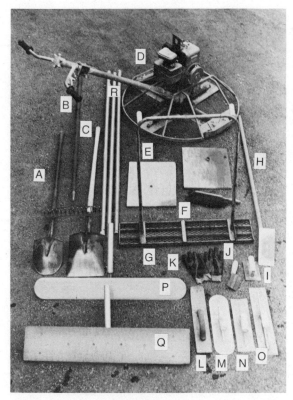

freshly placed concrete to smooth and flatten it out. The bottom side of the screed will determine the top of the slab.

Working the screed board in a seesaw motion across the slab works best. The motion is as if you are trying to cut the concrete in half while pulling the board toward you. This action flattens and levels the slab, and it helps to push down any rock that might be sticking up. Screeding is an integral part of concrete pouring. I, along with many other finishers, feel that if a job has been properly screeded, the majority of work has been done. Take your time and do an excellent job of screeding. A flat and even slab depends on it.

TAMP

The tamp is a metal tool. A heavy gauged wire screen, 3 feet long and 8 inches wide, lays on the bottom. Attached to it are two metal arms that run up and connect to a handle. The tamp is raised above the concrete for a few inches and then forced down onto the slab. The pressure pushes down the rock and brings up the cream. The wire mesh is tight enough to prevent rocks from going through, but it is wide enough to allow cream.

Tamping should be started immediately after

Fig. 6-1. All tools needed to finish concrete (A) round-point shovel, (B) rake, (C) square-point shovel, (D) finishing machine (optional), (E) knee boards, (F) foxtail broom, (G) tamp, (H) walking edger, (I) hand edger, (J) seamer, (K) rubber gloves, (L) hand float, (M) pool trowel, (N) scrubber and leaner trowel, (O) finishing trowel, (P) rounded fresno, (Q) bull float, (R) extensions.

shock absorber for the load. If you don't have one of these wheelbarrows, obtain one for the pour (Fig. 6-2).

Wheeling concrete is very hard work. If you can manage any way for the truck to get directly to the job site, do it. Don't get carried away and tear up half your yard. If all it will take is the removal of a small section of fence, though, go for it. If the job is a big one (more than 5 yards), don't forget about the concrete pump.

SCREED BOARDS

Screed boards are nothing more than straight 2 × 4 lumber. These boards are laid across and supported by the forms. They are pulled across the top of the

Fig. 6-2. For concrete work, you need a heavy-duty contractor's wheelbarrow with an air-filled tire.

the concrete is on the ground. Ideally, while some helpers are still screeding, another one can start tamping. After the concrete is on the ground, in the forms, and screeded, don't fool around talking to the driver or your friends. Waiting too long to tamp will cause you plenty of hard work. Stiff concrete is tough to tamp. You might have to really force down on the tool to bring up the least bit of moisture to say nothing of forcing down the rock. By getting right on it, you won't have to exert much pressure on the tamp. Wet concrete only requires a small bit of pressure; possibly the weight of the tamp alone would be enough.

While tamping, try to keep the bottom surface flat. Using it at an angle will cause waves and divets, making floating difficult.

All jobs should be tamped with two exceptions. First, a pea gravel mix does not have to be tamped. The screeding motion is usually enough to force down any small rocks. Secondly, never tamp a job that will be finished as an exposed aggregate. With an exposed slab, you'll want the gravel close to the surface. A good screeding and bull float will ensure an even surface.

Tamping is a very messy chore. Every time the tamp is dropped on the concrete, mud splashes for feet. It will get all over you, the house, a nearby wall, and anything nearby. Use tape to secure plastic sheets or newspapers against the house and any other thing you want to protect. Be sure that the bottom of the plastic or newspaper will not be so low as to enter the slab. You should only have to protect the house or wall up to 3 feet. If concrete splatters more than that, you are tamping too hard. These protective covers can be removed while you are out on the slab finishing. At that time, any concrete residue that falls off can be easily finished into the concrete.

BULL FLOAT

After the slab has been tamped, it should be bull floated. The bull float is a piece of smooth wood with a length of 3 feet, a width of about 8 inches, and a thickness of 1 inch. A special adapter is attached to the wood. The adapter is designed to accepted extensions (poles), which enables the user to reach the furthest part of the slab (Fig. 6-3).

The bull float is made to be pushed and pulled across wet concrete. It removes the bumps caused by the tamp and flattens the slab to a smooth texture. The weight and texture of the wood pushes down any rocks that may still be lingering. Its main purpose is to smoothe the cream and fill in any *cat's eye*, which refers to any small and shallow holes created by the tamp or uneven screeding.

The bull float can be maneuvered over the slab in any direction. Many times you will have to operate it from both sides of the slab and at angles to push excess cream into the shallow cat's eyes. Take care to produce a smooth surface.

Slight ridges will be left behind by the squared edges of the float. This is normal. If the concrete remains wet, you can float a second time. Once the mud begins to stiffen, though, the fresno will have to be used to further smooth the surface.

FRESNO

The fresno is a finishing tool that resembles the bull float. The basic difference is that it is made of steel as opposed to wood (Fig. 6-4). Its function is to get the slab as smooth as possible without getting out on it. A squared edge fresno will do a good job, but the rounded style is much better. Rounded edges will not produce any leftover lines or ridges. Thus, there remains less work in the final finish. I would recommend using the round fresno over the square one anytime.

Fresno use is determined by slab setup. The harder the concrete gets, the more often the fresno should be worked. Wet concrete should not be fresnoed more than twice. Overworked concrete will result in a lot of sand and grit at the top. This will make finishing more difficult. The same extensions will be used for the fresno as for the bull float.

EXTENSIONS

Lightweight 6-foot long metal poles are used as extensions for the bull float, fresno, seamer, edger, and broom. Each tool must have the adapter that mates to it. There are different brands of extensions, each with their own special way of attach-

Fig. 6-3. Extension poles and adapter on a bull float. Adapters on all tools should match extensions.

ment. When renting or buying such tools, be sure all of them are compatible with the extensions.

HAND TROWELS

Wood hand trowels are also called hand floats. They are used to lay down the concrete. During the pouring phase, hand floats are great for getting concrete into stubborn areas. Filling in cream along the side of a block wall and along rough surfaced foundations are just two of the functions. The hand

trowel is a handy tool and is not a real finishing tool.

Hand finishing trowels are made of steel just like the fresno (Fig. 6-5). They range in size from 4 inches wide and 8 inches long, to 6 inches wide and 18 inches long. I have found that having a smaller trowel for scrubbing and a larger one for finishing works great. The smaller one measures about 4½ inches wide and 14 inches long, while the larger one is still 4½ inches wide but closer to 16 inches long. The shorter trowel works better when using it to

Fig. 6-4. Round and square fresnoes with adapters.

roughen up the slab. One or two swipes with the big one will finish that area.

The blades on these trowels must be in good condition. Any nicks or dents will scar the slab. When renting your finishing trowels, be sure the edges are smooth and the bottom is free from defects. Do not slide your finger over the sides of the finishing trowels. After much use over concrete, they will get very sharp. A friend of mine slipped with his trowel one day and put a gash across his thumb that took eight stitches to close up.

If you are renting trowels, get two trowels for each finisher—one trowel for scrubbing and one for

finishing. Also, you may need a trowel to lean on. Stretching out over a slab requires a little help. A leaner trowel will enable you to reach out further and finish more of the slab from the edges without having to get out on it.

POOL TROWEL

The pool trowel is very similar to a normal steel hand finishing trowel and has rounded edges. The rounded features greatly reduce the amount of lines made by trowels. With this tool, you can finish a slab sooner, quicker, and easier. As soon as I had seen one in use, I went out and got one. They are great and will be perfect for the novice concrete finisher.

As with any tool, keep your trowels and finishing tools clean. The best way I have found is to fill a 5-gallon bucket with water (Fig. 6-6). After each use, the trowel can be quickly washed and hung on the side to dry. They also make a handy carrying case, too.

KNEE BOARDS

You must use knee boards to reach the furthest part

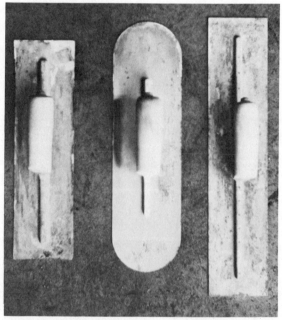

Fig. 6-5. Steel finishing trowels.

Fig. 6-6. A five-gallon bucket filled with water is an ideal way to clean hand tools during the pour.

of the slab with your hand trowels (Fig. 6-7). These items—two for each finisher—will be used as stepping and kneeling blocks on the slab. Measuring about 12 inches by 18 inches, you can easily maneuver them leap frog style across the slab. You simply place one in front of you, step on it, place the next one ahead of the first, step on it, etc.

Knee boards will leave an impression on the concrete. Make them out of very smooth plywood. The impressions can be roughed up with the scrubber trowel and then smoothed by the finishing trowel. The knee boards make a very small mess when compared to the deep marks left by just walking on a slab. The large flat area of the board evenly disperses the weight over a greater area than your foot. That is the reason for a lesser mar.

The key to good knee board use is to move them the least amount of times as necessary. Finish as much as possible from the sides—outside of the forms. Position yourself so that you can reach as much as possible from both sides of the knee board.

EDGERS

The edges of the concrete are those areas bordering

the main body of the slab: those parts that either touch a form, a stringer, or an expansion joint. If you're still confused, go out and look at your city sidewalk. You'll notice that the edge of the concrete is actually rounded off rather than left untouched and square.

An edger is made especially for rounding edges. The tool is made of steel, has a handle, and is similar to the steel finishing hand trowel. The variations from the hand trowel are that it is usually smaller, and one complete side is curved down. The arc of this rounded side can be as tight as ¼ inch or as wide as ½ inch, with ⅜ inch seemingly the most popular (Fig. 6-8).

Walking edgers are also available. Equipped with an adapter, walking edgers can be attached to the end of an extension. That feature lets you walk along and effect the edge rather than having to get on your hands and knees to do it. Walking edgers are generally a little wider than hand edgers. The

Fig. 6-7. Knee board with attached handle.

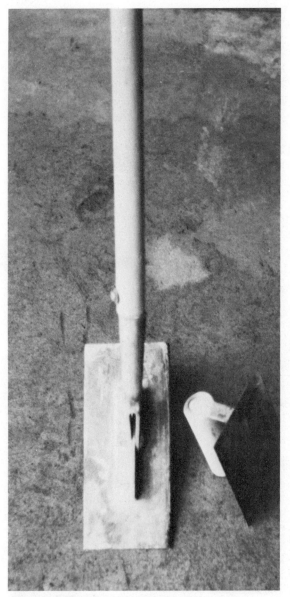

Fig. 6-8. Hand edger and a walking edger.

bull float. The edge will not stay nice and tight because the concrete will still be wet. The fresno will also fill in the sides with cream. By edging early, all of the rocks are gone and out of the way. When it comes time for the final edge, you have nothing but cream to work with.

SEAMER

The seamer, a specialized tool designed to install control joints, will not be necessary for large pours such as patios. The purpose of a control joint is similar to an expansion joint. Rather than actually separating the concrete, a seamer puts a crease in the surface. If all works well, instead of the concrete cracking across the center of a section, it will be controlled and crack inside the seam.

The seamer tool is made of metal and has a raised ridge down the middle (Fig. 6-9). To effect

Fig. 6-9. Hand seamer.

wider body helps to make a cleaner edge with no unsightly ridge left over.

If you decide to use a walking edger along with a hand-held one, make sure that each has the same arc. By using two edgers with different arcs, you will be fighting yourself the whole day.

I try to put on the first edge right after the first

the seam, a straight piece of 2 × 4 lumber is placed across the concrete. One side of the board will be used as a guide. With one edge of the seamer resting against the 2 × 4, the tool will be assured of going in a straight line.

The initial seam should be made after the first bull float. Like the first edging, the seam will not stay perfect. It will get filled up with cream from subsequent fresno application. After the first seam, though, you might be able to use a walking seamer and not need the 2 × 4 guide (Fig. 6-10). Since the

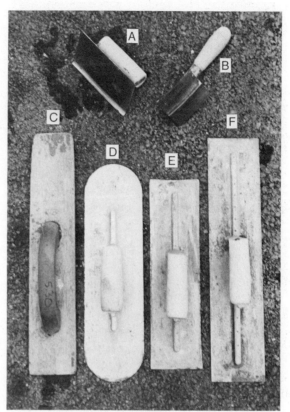

Fig. 6-11. Hand finishing tools: (A) hand edger, (B) hand seamer, (C) hand float, (D) round pool trowel, (E) scrubber trowel, (F) long finishing trowel.

groove will have been already established and the rocks out of the way, the seamer should follow its own path and renew the joint.

Seams are used mainly on walkways. With expansion joints, seams take up the slack between them. If you looked at your city sidewalk, you'll see an expansion joint with felt every 10 to 12 feet. Between them, at about 3 to 4-foot increments, will be a seam. Expansion joints and seams work together to prevent unnecessary and undesired cracks (Fig. 6-11).

BROOM

Putting on a broom finish is easy. This finish may hide finishing errors, too. Only a soft-bristled push broom will work. I have had good use out of a heavy-duty, horsehair push broom. The wide fea-

Fig. 6-10. Walking edger. Notice the depth of the blade and the width of the tool. These features will make an excellent seam.

ture enables me to cover a large area with each swipe, and its construction holds up time after time. Special concrete brooms are available, but they are not needed if you have a good push broom.

The tiny lines left by the soft bristles scratch the surface just enough to roughen it up and give traction. In the process, small trowel lines are wiped off. To reach sections of the slab many feet from the edge, you can place an extension handle over the broom handle. Secure the two together by attaching a small wood screw through the extension's coupling hole and into the broom's wood handle (Fig. 6-12). Other extensions can be linked with the first to give you the reach needed.

On very long and wide slabs, you may have to broom one-half of the slab from one side. Broom the other half from the opposite side.

Broom lines do not have to be straight. By wriggling the broom while pulling it toward you, you can make wavy lines on the slab. Another design can be created with a short, hand-held foxtail broom. While out on the slab with your knee boards and trowels, lightly broom the surface after you have trowel finished. Keeping your arm still, broom sections by simply moving your wrist in a half-moon sequence. This is called the windshield wiper broom finish. The effect will look like dozens of windshield wiper motions over the slab. You can continue all the motions in the same direction, or you can do one section in a forward direction and the next in a backward or reverse direction.

Before applying the broom finish, the concrete should be set up and finished. If the concrete has really set up on you and the bristles are not making much of a dent in the surface, soak the broom with water and push and pull it over the concrete. With wetter concrete, set the broom down at the furthest point away from you and pull the broom to you. On harder concrete, you may have to push and pull the broom across the surface.

ROCK SALT

If a rock salt finish is in your plans, you need to buy some rock salt. It is available at most supermarkets and is labeled as water conditioner salt. I frequently use the Morton coarse brand of salt. The name

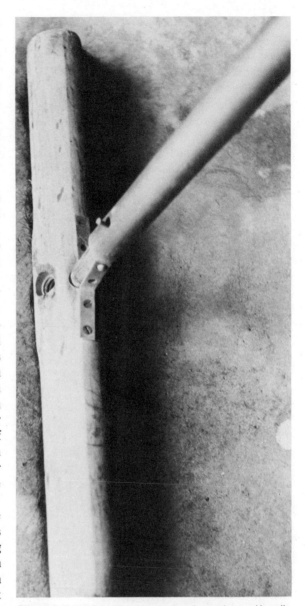

Fig. 6-12. Extension pole placed over a broom's wood handle and secured with a wood screw through the hole in the extension and screwed into the wood handle.

brand isn't an important feature, but the granule size is. Granules too large will leave holes that are just too big. Smaller granules will not leave the desired impression. The granules you should use are about three-fourths to the full size of a common pea.

When the slab has been completely hand troweled and finished, run a light broom finish over the top. After that, evenly spread the rock salt over the entire surface. A heavy application is best because not all granules will penetrate. Use your knee boards to get back out on the slab. Use a hand trowel to pound in the salt. The knee boards should not mar the surface. For the most part, they will rest directly on the salt.

ETCHING

The etching design is put on after the concrete has been troweled to a finish and broomed. It must be well set up before you begin. The surface should be so hard that your knee boards will not make any impressions while you are on them.

The only tool needed to do the etching is a common 1-inch-wide wire brush. These are available at hardware stores and lumberyards. The type I use is commonly used by painters to clean brushes. It is 1 inch wide and about 4½ to 5 inches long. A handle is attached to make using it easy.

Every edge, seam, and expansion joint should be etched. Follow the forms straight the entire distance, making only a 1-inch-wide scribe. In the

Fig. 6-13. Heavy-duty rubber boots.

corners, round off the intersection. After all the borders and reachable seams and expansion joints are done, you can get your knee boards and start in on the main slab. Other pictures in this book shows etched slabs. You can use them as a guide. For the most part, however, the design is free-formed.

Attempt to make each stepping stone about the same size. This will give your slab a more even look. Don't forget to round off each and every intersecting point. Putting on this finish is fun. You will enjoy looking at your artistic talents afterward. By the way, don't be worried about the concrete crumbs on the slab as a result of the etching. The main surface will be hard enough to resist them, and they will wash off the next day. Don't set your knee boards directly on them. There is no sense in trying to push them into the slab.

The wire brush is also handy for cleanup projects. Stubborn concrete on the forms is easily removed with a wire brush. Fresh concrete is also easily brushed away from walls and other existing concrete.

RUBBER BOOTS AND GLOVES

Cement powder mixed with concrete is caustic and will ruin shoes. Wear rubber boots during the pouring phase (Fig. 6-13). It will not be necessary to wear them all day—just when you are screeding, tamping, and maybe during cleanup to keep your feet dry.

Wear rubbers gloves, too. During the pouring and screeding, you will have to reach into the mud for different reasons. The lye in the concrete will tear up your hands. They will dry out, crack, and possibly peel. Many professionals always wear rubber gloves.

POWER FINISHING MACHINE

As shown in Fig. 6-1, a gasoline-driven engine is used on a concrete finishing machine. The two-stroke or four-stroke engine propels four finishing blades at the base of the machine. The rotating speed of the blades is determined by the speed of the engine. The angle of the blades is adjusted by the round knob in the center of the handles.

These machines are great finishers, especially for very large jobs, A person should have ample experience with one before using it. For most slabs you would lay around your home, such a machine is not necessary. Timing with the power finisher is critical. For the inexperienced user, I do not recommend its use. If you have a friend that has used the machine and you want to try it, you can rent one.

Chapter 7

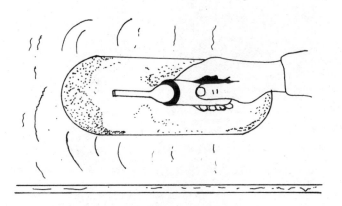

Preparations

PREPARING THE JOB SITE FOR FORMING IN-cludes moving things out of the way. Patio fur-niture, planter boxes, swing sets, and the like should be moved to another part of the yard. You will want plenty of room to work. Some forms may be long, and you will not want to knock anything over and break it. Neither will you want to trip over something and hurt yourself.

REMOVING OLD CONCRETE

Many houses have a small concrete stoop at the front or rear porch area. If your job will be poured around it, you might consider removing it and pouring a fresh slab. Some people have laid new patio slabs around this small stoop, and the job looked tacky. I have always broken them out to make room for a complete slab (Fig. 7-1).

Dig away some ground from beneath the step. Digging out the surrounding edge will help to break loose the step from next to the house (Fig. 7-2). After enough dirt has been removed, use a heavy-duty pick to pull up the step and lift it up (Fig.

7-3). Use caution as to not hurt yourself. Concrete is heavy. The reason behind loosening the step is so that it will be easier to break into pieces and haul away. If the pick doesn't give you the leverage needed, use a long 4 × 4 piece of wood with a block under it (Fig. 7-4).

You can use the pick to pry the step away from the house. Do this by inserting the pick between the edge of the loosened step and the house. Using the pick for leverage, try to scoot the step away. By doing this, the step will be moved from the house enough so when you begin to break the concrete with a sledgehammer, no damage to the house will occur.

A heavy-duty sledgehammer, at least an 8-pound one can be used to hit the concrete and create cracks. Continue to hit the surface along cracks until the crack is wide and seems to go through the step. When definite cracks are present, use the pick to pry up the broken chunks (Fig. 7-5). Afterward, you can haul the chunks off to the dump or use for fill as needed.

Fig. 7-1. Removing this small front porch will make the entire porch and entryway look professional and not pieced together.

OUTLINE THE JOB

The simplest way to begin the forming job is to place the forms in their approximate locations. This will form an outline around the job and get the material close to its proper place (Fig. 7-6). This practice will help you determine which forms should be laid at each point. Use the forms intelligently. In other words, don't use a 14-foot form when only a 10-foot one is needed. Do not waste forms. Measure each one and place it in the location that best suits its length (Fig. 7-7).

After placing the forms in their specific locations, lay out the stakes. Place a stake on the ground near the forms every 3 feet. The stakes will be easy to reach once you begin placing them.

Use the forms to their greatest potential. If you have one side of the slab that measures 13 feet, use a 14-foot board. At the corners, some forms can extend past the slab (Fig. 7-8). The extra 1-foot extension will not hurt anything, and it will save wasting a nice 14-foot form.

On jobs where a short 1 or 2-foot form is needed, cut a piece from a short form. Don't steal a piece from a nice, long 16-foot form. Use it from a piece of scrap—a 3 or 4-footer.

MAKING ROOM FOR FORMS

Some jobs require removal of ground for the forms to fit. For example, if you are planning to lay a slab in an area that was previously filled with decorative

Fig. 7-2. First, dig out dirt from under the front of the stoop. This will help loosen it from its base.

bark, and that space had been previously dug out compared to the surrounding landscape, the forms might not fit into the right space. Instead of digging out too much dirt, lay the form down in the general area and use it as a guide for digging (Fig. 7-9). You won't be digging out more than needed, and you'll save some work. Another benefit is that your yard will suffer only a minimal amount of disruption. A square-point shovel works good with this type of digging (Fig. 7-10).

BLOCKING FORMS TOGETHER

On slabs with long runs and some walkways, it is much easier to form when long forms are used. Lumber in lengths over 18 to 20 feet is very hard to find in straight condition. It is simpler to block two forms together using shorter and straight lumber. On a run of 24 feet, I would use two 12-foot forms. Before placing them, I would lay them on a flat surface and block them together. Laying flat, you can join them together and ensure they are in line. Butted end to end, use a string or another long form to ensure the tops are straightly matched. Make a note as to which 2-inch side will be the top of the form. By doing this, you can be sure to put the block on the correct side.

To join them, use a short piece of 2 × 4, preferably about 1½ feet to 2 feet. Lay the block on top of the joint and securely nail it to both of the forms (Fig. 7-11). You will need some help in mov-

Fig. 7-3. Using a heavy-duty pick, pry up the front of the stoop to break it loose.

ing the new form. The joint will be strong enough as a form, but not as a board that will not bow in the middle.

ESTABLISHING HIGH POINT

Deciding on the high point of the slab is critical. Water runoff is a very important concern to many homeowners. You must determine where you want water runoff to go. Always remember, though, that no water should ever be allowed to flow toward or puddle next to the house or any other structure.

When pouring concrete next to the front or rear door, you can use the existing doorsill as a guide (Fig. 7-12). By making a mark on the house level with the sill, you can measure down from it to the point where you will want the top of the concrete to go. I like to stay at least 2 inches below any floor level. This ensures that no water will get into the house when I wash if off. When pouring a front porch and walkway, this high point marker can be

made on both sides of the front door (Fig. 7-13). These marks can be joined by a chalk line. The chalk line can also be made to mark that entire side, giving a very good reference for grade (Fig. 7-14).

Chalking a long line on the house might take three people. The line holders on each end should stretch the line tight. If both of them have marks to go by, the line should be resting on their mark. If the reference mark is a short line in the middle, the middle man will have to direct each of the end men whether to raise or lower their line. Once the line is stretched in its proper position, the middle man can pull the line away and let it snap against the house (Fig. 7-15). A straight and very visible chalk line will be made, and your grade reference will be in place.

SQUARING

To square up a side form, use the square tool against the house and side form. The far end of the form will have to be moved for it to match the square tool and the house at the same time.

To set a slab square that will not be against the house or other reference, you'll have to use a square and string. For instance, if you wanted to pour a small slab in the backyard 10 or 20 feet from the back of your house, you'll need some kind of a reference to keep it squared with the back wall of the house. To do this, place a stake right next to the back wall at a point where you will want one side of the new slab to be. Attach one end of the string to it. Place another stake past the far end of the proposed slab; the string will stretch from the back wall of the house and about 2 feet past the far end of the job site.

With the string attached to the far stake, go back to the end next to the house. At that point, use the square tool to check the string against the wall of the house. You will have to move the far stake a few times until the string becomes perfectly square with the back wall. When that string is square and still attached to both stakes, measure a distance from the first stake and equal distance to the length of the proposed slab. Do this against the back wall of the house. At that point, place another stake. Do the same measuring for the far stake and place a stake at the same distance away from it. Attach a

Fig. 7-4. If a pick does not give you enough leverage, use a 4 × 4 and a block under it as a lever.

Fig. 7-5. After cracking the concrete with a sledge hammer, use a pick to pry out the chunks of broken concrete.

Fig. 7-6. Outline the job first. This will give you the chance to properly place lumber without having to cut long forms.

Fig. 7-7. Always measure forms before you put them in place. Sometimes boards measure longer than what they are supposed to be. A 10-foot board might really measure 10 feet 2 inches.

string to both of those stakes and be sure of the square.

Both of those strings will ensure that each side of the slab will be square with the back wall of the house. To establish a square for the front and back sides of the slab, again use stakes and string. First, determine the front of the slab by measuring from the back wall of the house. Place a stake at that point on either side of the new slab area. Attach a string to it. Place another stake on the opposite side of the first. Use a square tool to ensure a square corner. Do the same procedure for the backside of the slab. The strings will be your guides in placing the forms. Once the first form is set and staked in place, you can use it as your guide (Fig. 7-16).

OTHER TASKS

Before delving into the forming process, there are a few other things to remember. Besides preparing a way for the concrete to get to the forms from the truck, you should consider routes for wheelbar-

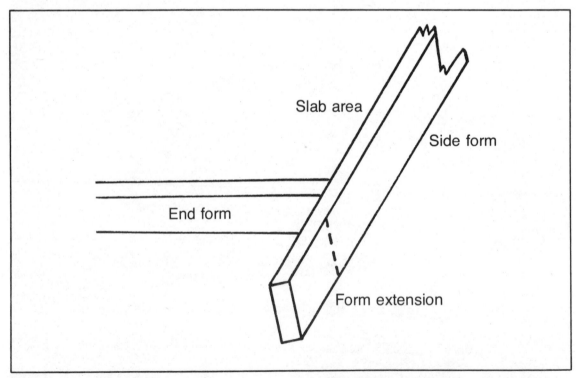

Fig. 7-8. When using forms longer than needed, a short extension past the corner will pose no problem and prevent the need to cut it off.

88

Fig. 7-10. A square-point shovel works best because it will square off the dirt at the base and pull away the side.

Fig. 7-9. Using the square-point shovel, square off the sides of the job to fit forms. A clean outline will prevent unnecessary damage to the landscape.

Fig. 7-11. A 2 × 4 block used to join two forms on a long run.

Fig. 7-12. Using the door sill as a level guide.

Fig. 7-14. Chalk line snapped against house foundation as a grade and concrete reference.

Fig. 7-13. Marking the high point on each side of the door. Mark out an equal distance from the door so the side forms will be equally set.

Fig. 7-15. Three men working together to snap a long chalk line.

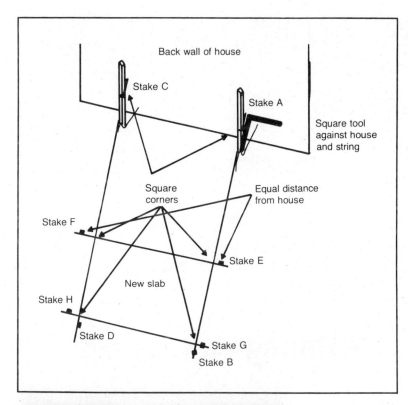

Back wall of house

Stake C

Stake A

Square tool
against house
and string

Square
corners

Equal distance
from house

Stake F

Stake E

New slab

Stake H

Stake D

Stake G

Stake B

Fig. 7-16. Using a square, string, and stakes will help you set up a slab square with the house. Stakes A and B, C and D, E and F, and G and H go together.

Fig. 7-17. Make solid heavy-duty ramps for wheelbarrows when they will be wheeled over forms. This will prevent the forms from moving.

rows. If there are obstructions in the way, ramps must be built for the wheelbarrows to roll over. In any case, at least one ramp must be provided for the wheelbarrow to get over a form and into the pouring area (Fig. 7-17).

Upright plumbing cleanouts must also be provided. I have had good luck using coffee cans over them during the pour. When the concrete has cured (usually after two weeks), I cut the bottom off the can and expose the cleanout. There are special covers made for them, but the coffee cans work fine. The covers I refer to are those commonly used to cover water meters at the front of your house.

Horizontal vents for the area under the house floor should also be left open. One-by-four forms can be used to form an open area around them, and the forms can be left in place after the pour.

You should also make provisions for water and electrical pipes. If in the future you might want water or electricity to an area near the slab, a plastic pipe sleeve will provide you with that option later. You won't have to tunnel under the slab.

Chapter 8

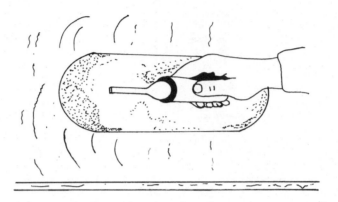

Forming

FORMING HAS ONE SIMPLE CONCEPT—TO KEEP the concrete in one place and one particular design. Slope is a very important concern with all outdoor slabs. Your forming job must be strong enough to keep the concrete in place, and it must allow for proper water drainage. One other facet that is not as critical, but just as important, is straightness. Straight forms will provide a straight concrete edge. That feature will make your slab look like it was done by a professional.

FIRST FORM

The examples I will be depicting are slabs that come off the end of a house. The wall on the house will serve as a reference. The first form set will be one of the sides.

After a mark has been made on the house showing the top level of the concrete, a side form can be butted to it. Be sure the top of the form will meet that mark. If the grade is too deep for the form, fill in with dirt. If the grade is too high, dig out dirt until the form will fit. Lay the form in its designated position. Use the square to help place it.

Then pound a stake into the ground next to the form about 1 foot from the house (Fig. 8-1). With the stake in position, lay the form next to it. Brace the board so that it will stay next to the stake. Go to the other end of the form and be sure it is steady. Place a duplex nail through the first stake and into the form. While nailing, be sure the top of the form matches the mark on the house. Use the square to be certain the form is perpendicular to the house. If the form is set, proceed to place a stake next to the form at the far end (Fig. 8-2).

At this time, place the level on top of and in the middle of the form (Fig. 8-3). If the slope is way off, get the shovel and dig away, or fill in, as needed. The bubble inside the level will rise to the high side. Therefore, it should be past the first line toward the house (Fig. 8-4). If the bubble is close, go ahead and nail the second stake to the form. With both stakes in place and nailed, the form should be lightly supported.

Although the level might not be perfect, it should be close. Then you can begin setting the other stakes. You will be attempting to set the form

Fig. 8-1. Place the first stake about 1 foot from the house.

straight. An easy and accurate way to do this is with a string. At both ends of the form, place a small nail into the top inside corner of the form. Attach the string to these nails. Pull the string tight and secure it. The string should run flush against the form at both ends.

When tight, the string will be your straight-edge guide (Fig. 8-5). You can go back to the front of the form and begin placing the middle stakes. Push or pull the form until the edge touches the string. When it is in place, pound in the stake. If the level of the form was perfect, and it is not a bad idea to recheck it here, you can go ahead and nail each stake as you go. If you will have to do even the

slightest bit of adjusting, don't nail the stakes yet (Fig. 8-6).

Placing the middle stakes can sometimes be difficult. If the form is bowed, you'll have to put pressure on the stake and allow the stake to push the form, too (Fig. 8-7). The best way for me is to stand on the outside of the form, hold the form and the stake together with my left hand, and begin driving the stake. When the stake has started to penetrate the ground, I hold the form with my left hand and brace my left foot against the stake. You can swap sides as needed. You can also stand on the inside of the form if it is bowed in, and you must put pressure on it to push it out. This is best done with your foot.

Setting stakes can get tricky. You may have to use steel stakes in rocky areas. You'll have to use

Fig. 8-2. Place the second stake at the other end of the form.

93

Fig. 8-3. Check the slope by placing the level on top of the middle of the form.

long wooden stakes in soft ground areas. On special occasions that require the need for footings, staking is very difficult unless you take this word of advice. In Fig. 8-7, the former is attempting to form a job where the footings have already been dug. He must use very long stakes to reach down far enough to get support. Every time he goes to sink a stake, the dirt inside the footing falls away from the side and leaves the form loose.

I form the job first and then dig the footing. I don't have the problems of collapsing dirt, and I won't miscalculate the line for the footing. With the forms in place, I can dig a perfect footing straight down from the forms and never disturb a stake. Forming is much easier all around. Laying out the forms on flat ground provides a good solid base. I

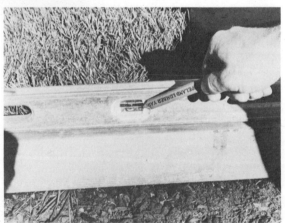

Fig. 8-4. You have a good slope when the bubble touches the second line, or one-eighth of the bubble is past the line toward the high side.

Fig. 8-5. Use a string attached to both ends of the form as a straightedge guide.

don't have to bridge the forms to keep them in place while staking (Fig. 8-8).

Let's go back to the first form. If the level is correct, nail the stakes as you go along. If the level is off, don't nail. With only the end stakes nailed, you can move the form down quite easily. Moving it up would require removing the nails. To place one end of the form down slightly, all you have to do is pound down on the stake (Fig. 8-9). The nail through the stake and the form will cause the form to move right along with the stake. Keep an eye on the level mark and stop when the top of the form lines up with it.

A slight problem you may encounter is when one end of the form moves off its side mark. That measurement will determine the width of the slab.

If the form moves less than an inch from the mark, you can force it back in place by extra staking. Driving a stake in at an angle can force the end of the form to move slightly (Fig. 8-10). This is not a good habit to get into because it could throw the entire form out of whack.

Before going onto other forms, the first one should be completely set and properly staked. You can never use too many stakes. Figure 8-11 shows a first form setup. The extra stakes lying on the ground next to it are to be used as kickers. As you can see, this concrete former is making absolutely sure his forms won't move (Fig. 8-11).

Because of the grade, some slabs will have a side that rises a few inches more than the other. On the low side, the top of the concrete will have to be

Fig. 8-6. Don't nail stakes until the form is set.

× 6 or 2 × 8. After the entire job is formed and you begin grading, you can leave a space 4 inches to 6 inches away from the inside of the form toward the fill. Because of a deeper than 4-inch base, that area will have to be filled in with dirt. By leaving a trough next to the form, you will be guarding against future erosion under the slab.

SECOND FORM

Generally, the second form I place on a slab that has the wall of a house as a guide is the other side. I save the end form for last.

Fig. 8-7. Using pressure on the stake to push the form into place.

more than 4 inches off the ground. In those cases, form the first stake as you would any other. After the staking and nailing is done, nail an additional form under the first (Fig. 8-12).

For sides that are totally low the entire length of the form, you might have to use a wider form—a 2

Fig. 8-8. Laying stakes across the footing will give the form support and keep it in place.

Fig. 8-9. To lower the form, simply pound on the nailed stake until the form reaches its mark.

outside of the first form and stretch it out to make the mark for the second form, the mark should be made at 20 feet, 1½ inches. The 1½ inches will compensate for the width of the 2 × 4.

Fig. 8-10. Kicker stakes help to force form ends into place.

I measure and mark a point on the house (near the ground close to the form) the exact distance that equals the width of the slab. For example, if the slab is going to be 10 feet wide and 20 feet long, mark 20 feet from the inside of the first form. Remember that the width of the 2 × 4 form is really 1½ inches. Therefore, if you attach your tape measure to the

You should have two marks on the house—one for the width and the other for the level. Using those marks, form the board just like the first. Place the first stake, brace the form, place the second stake, check the level, and nail. Again, use the string for a straightedge guide (Fig. 8-13). Drive the middle stakes into position and nail them only if the level is correct. When nailing stakes, stand on the inside of the form, brace your foot so that it pinches the form between it and the stake, then nail (Fig. 8-14).

For added support, kicker stakes can be added to the form. These stakes are placed after the others. They are set at an angle and prevent the top of the form from moving (Fig. 8-15). With kickers in

Fig. 8-12. To secure concrete on sides deeper than 4 inches, place an additional form under the first.

place on a straight form, you can feel very secure in knowing that the form will not move and that your slab's edge will be perfect.

END FORM

The end form is the one that runs parallel with the wall of the house. It will connect both of the far ends of the first and second forms. Setting long end forms is not difficult if you follow some guidelines. For short runs, you can simply nail each end of the form to the ends of the side forms (Fig. 8-16). This will give you a brace on each end, providing both side forms are of the proper length. Sometimes lumber is not cut exactly to length. A 10-foot board might actually measure 10 feet, 2 inches. Check your forms before installing them.

With both ends of the end form nailed to the side forms, you can stretch a string and stake it in place. Check the level for proper slope. If the width of the slab has the slope and you want the length to be level, the end form should be level (Fig. 8-17). If it is not, you'll have to do some readjusting. Nine times out of 10, though, the end form will have a slight slope, too. This will cause water to run to one corner of the slab. At that corner, you might even want to install a drain to further alleviate any runoff problems.

Fig. 8-11. Use plenty of stakes. Laid out here are the kicker stakes.

Fig. 8-13. On the second form, use a string as a straightedge guide.

Fig. 8-14. Use your foot as a brace against the form when nailing stakes.

Fig. 8-15. Inserting a kicker stake in the middle of the second form.

To be doubly certain of the dimensions, check the square at each of the far corners (Fig. 8-18). Because the top of the side forms was square when you placed them, the chances are good that they will be square at the bottom. For most patio and porch slabs, this is not a major problem. If the square is off by only a fraction, there should be nothing to worry about.

End forms that are longer than 14 feet and those that require you to join two forms together can be formed a little differently. Instead of relying on a string, you can use an additional board that has the same identical measurement as the side forms. Use it as a block.

By placing the block between the house and the spot you want to place a stake, the block will serve as a brace. Being the same length as the side forms, you will not have to worry about the form being set out too far or in too close. Using the block makes staking much easier. By placing the stake next to the form, you can brace your foot against the stake, which puts pressure on the form supported by the block. The end of the block is resting squarely on the house, so you have a perfectly strong support.

KICKER STAKES

Kicker stakes add a lot of support to forms. They ensure the top edge of the form being straight and not bowing under the pressure of the wet concrete (Fig. 8-19). Kickers are an absolute must when forming jobs with footings (Fig. 8-20). The regular stakes will do their part even though a lot of dirt has been removed in front of them. Their support comes from the concrete. When pouring a slab with footings, the footings are filled first. The concrete should not be poured to the top of the form until the footings have all been filled to grade level. Therefore, the concrete in the footing will give support to those stakes. You don't ever want a form to break away. It is always best to be safe than sorry, and that's where the kickers come in.

Tall forms for steps or on uneven grades that require a form wider than 4 inches should always be heavily reinforced with additional staking (Fig. 8-21). Due to the amount of increased weight from the concrete, taller forms should be staked no more than 2 feet apart. Extra long stakes and plenty of kickers are a must,

GRADING

Grading the base for a concrete job will level the ground to its proper depth. A solid and even grade will ensure an accurate concrete yardage estimate and will prevent cracks.

Cracks can appear in slabs that are placed on an uneven base. If the concrete is 5 inches deep in one

Fig. 8-16. Nailing the end form to the ends of the side forms already placed and staked.

place and only 3 inches in another, there will be a weak spot. A crack will soon appear at that weak point.

Many concrete finishers consider grading the hardest job in the business. It can be, especially if you go about it haphazardly. There is no sense in digging out too much dirt when you'll just have to put it back. Therefore, follow the next few grading tips and make your grading job an easier one.

The simplest way to determine the grade is by placing the screed board in place and measuring the depth beneath it (Fig. 8-22). The bottom of the screed board will determine the top of the slab. Having 3½ inches to 4 inches under the bottom side of the screed shows a good grade. If the ground is too high, you'll have to remove and level it. Sometimes you'll have to use your pick (Fig. 8-23). Use the pick end to break up the dirt and the blade to further clean it out. If the ground is only a few inches away from grade, don't dig a deep hole. Take short strokes and try to break lose only the amount of dirt needed.

After the ground has been broken up with the pick or if the dirt is soft enough, you can use a rake

Fig. 8-17. Before securing the form, make sure enough dirt is removed to allow the form to set at its proper level. Note the level tool sitting on the form.

Fig. 8-18. Use a square tool to check the square at the corner.

Fig. 8-19. Kicker stakes give support to the tops of the forms, ensuring a straight side.

to move the dirt (Fig. 8-24). A heavy-duty rake will move a lot of material and help level out at the same time. A quick way to grade with a rake is to measure its width and use it closely to the screed (Fig. 8-25). Each time the rake fits under the screed, you'll know just about what the grade is. The best way to do this is to operate the rake right alongside the screed. The rake is always next to the screed, giving you very close reference.

Another way to help grade is by attaching a grade screed to the bottom of the screed board (Fig. 8-26). To make this, simply attach two stakes to the screed and nail them to another 2 × 4 on the bottom. First, lay the screed board down on a flat surface. Second, get another 2 × 4 that is just a little shorter. The shorter length allows the grade screed to fit inside the forms while the main screed rides on top of them. Place the grade screed on its side and

103

at the bottom of the main screed. Using wood stakes, nail the stakes to each board in an upright manner. The bottom of the stakes should not go past the bottom of the grade screed. Because the bottom of the grade screed will be riding on top of the base, there will be no room for protruding stakes.

With the grade screed nailed to the stakes on the bottom and the main screed nailed to the stakes on the top, you're ready to go. Place the screed on the forms and grade the base so that the bottom screed board just barely touches the ground. This maneuver will absolutely ensure that your base is graded to 3½ inches to 4 inches every time.

One final way to determine grade is by using a string. Stretched out from the tops of the side forms, the base should always be from 3½ inches to 4 inches from the string (Fig. 8-27). This method is a little tougher because you cannot move the string as much or as easily as the screed board. You will have to use the tape measure to test the depth every so often. Along with the tape, you can mark 4 inches on a stake and use it as a guide (Fig. 8-28). With the stake, you can walk along the string and constantly check the depth.

Fig. 8-21. Tall forms that will support a lot of concrete must be heavily reinforced.

Fig. 8-20. Kicker stakes are an absolute must when forming around footings.

Fig. 8-22. Measure the depth under the screed to show the depth of the job.

SPECIAL FORMING

Sometimes you must be creative and form something a little unusual. A short run of steps leading to a pool or spa does not have to look ordinary. In Fig. 8-29, the former has used redwood stringers to outline a set of off-center steps. Leading from a deck to a spa, the center step was formed a little to the right. This forming job is also set up to keep the stringers in place after the pour. Later, they can be

stained to match other pieces of decorative wood on the house.

Some forms might bow toward the inside on a long run. To compensate for that, a single stake can be placed inside the form to push out the bow (Fig. 8-30). After the concrete has been laid, the stake can be pulled, letting the weight of the concrete hold it in place. This procedure is very common when using thin strips of benderboard to form curves.

Fig. 8-23. On hard humps of dirt, use the pick to loosen it up. Use the pick end to break up the clods and the blade to smooth out the area.

When joining thinner forms with 2 × 4 lumber, as with 1 × 4s used on curves, placing the stake a little differently can help make the job go quicker (Fig. 8-31). By driving the stake in sideways at the butt of a 2 × 4 form, the thinner wood will not stick out or be pushed in at the edge. If the edge did not line up with the edge of the 2 × 4, a lip would be made. The important feature in Fig. 8-31 is that both the top of the 2 × 4 and the top edge of the 1 × 4

line up. The stake can also be nailed to the 2 × 4 for added support.

When forming a job with a 4 × 4 fence post at the corner, it might be difficult to drive a stake down next to it. This is because most fence posts are sunk in concrete. Therefore, it might be easier to nail a stake to the 4 × 4 and then lay the form next to it (Fig. 8-32).

Pouring a walkway along the side of a bank can

Fig. 8-24. Rakes are excellent tools for grading.

cause unique problems. The biggest concern will be erosion. Heavy rains could cause the concrete to become eroded when the bank begins to wash away. One method to prevent such erosion is to dig a trough along the side of the form. Make it deep enough to secure itself. The trough only has to be 4 to 6 inches wide. For added security, you can place a few pieces of rebar into the area for anchors (Fig. 8-33). The rebar should be sunk into the ground at least 6 to 10 inches and be allowed within 1 inch of the top of the concrete.

When pouring a job that requires footings, you will probably have to include rebar, too. Hanging long rebar rods in the center of a footing is done with wire and stakes. First, stakes are laid across the footing, supported on one side by the form, and on the other by the ground. *Tie wire* is used to secure the rebar to the stake. After the concrete is in the footing, the tie wire can be cut and the stake removed (Fig. 8-34).

FORMING STEPS

It takes time, patience, and ingenuity to form steps. Get help and advice from an experienced step former before tackling a tricky step job. There usually is an extra amount of concrete that goes into steps. That factor alone will account for an added amount of concrete and a much greater amount of weight forced on the forms. The step facing form, the one that holds the actual step in place, must be removed before the concrete is hard so that it can be finished.

For very small steps that will act more like curbs, you simply have to place a form in the place that you want the face of the curb to be (Fig. 8-35). Stakes that support it are in the middle of the slab

Fig. 8-25. By measuring the height of the rake, you can tell how deep the grade is when the rake is used under a screed board.

107

Fig. 8-26. A grade screed consisting of a 2 × 4 board attached to the screed board helps to rake off the right amount of dirt.

and will have to be removed after the concrete has begun to set up. Holes left by the stakes will have to be patched and finished. In Fig. 8-35, the curb form will be left in place after the pour. This is demonstrated by the nails sticking out of the left side. Those nails will act as anchors in the concrete to keep the board in place. If the board was to be removed, the nails would not be there. After re-

moving the board, the concrete on the curb face will have to be finished.

Along the same lines, a normal step can also keep the form in place. Take two 2 × 6s and place one on top of the other to form a large step (Fig. 8-36). Staking with redwood stakes on the inside removes the chore of removing outside stakes after the pour. Again, the nails are used as anchors to

Fig. 8-27. A string can be stretched across the slab area to be used as a grade guide.

keep the board in place after the concrete has set up. The 2 × 4 you see attached to the form and going to the right is a stringer. It will be left in place as designated by the nails on the side.

Figure 8-37 shows a front view of Fig. 8-36. This will be the side of the step that will be visible—the other side will have concrete poured up to the top. The metal stake on the left-hand side of the form will have to be removed after the concrete has had a chance to begin setting up. Due to the amount of weight that will be pushing against this form from the concrete, an additional 2 × 4 board has been placed against the form next to the metal stake as a brace.

Sets of rising steps are the hardest of all to form. Basically, the top step should come out as far

Fig. 8-28. Placing a mark 4 inches from the end of a stake makes a good guide for measuring depth.

Fig. 8-29. Redwood forms that will stay in place after the pour and the offset design add flair to an otherwise normal step.

Fig. 8-30. Notice the stake inside the form to push out a bow.

Fig. 8-31. Stake placed sideways to take up the room not used by the 1 × 4 curved form. The support will keep the top edge of the 1 × 4 in line with the top edge of the 2 × 4.

110

Fig. 8-32. A stake nailed to a 4 × 4 post braces a form.

Fig. 8-33. Trough and rebar anchors along the deep edge of a walkway will prevent erosion of the concrete later.

Fig. 8-34. Stakes spanning the footing support tie wire that holds rebar reinforcing rods in place.

as possible and still be level with the grade at its level. The second step will come out level with the bottom of the first as far as it can go. The third should come out from the bottom of the second, level, and as far as it can go, and so on (Fig. 8-38). You can take a measurement on how far out (horizontally) the entire set of steps will come. Measure how high the entire set will have to be (vertically).

The vertical distance can be divided by factors of 7 inches and less, with 7 inches being the tallest step advisable. If the vertical rise is 28 inches, then four 7-inch steps would do the job. If the rise was 18 inches, then three 6-inch steps would be needed.

The horizontal factors determine the actual space on the surface of the step. You should allow a minimum of 10 inches of step surface for each step, and more would not be bad. If the horizontal distance was 30 inches, you would have room for three 10-inch-wide steps.

If the vertical distance is much greater than the

Fig. 8-35. This form will hold back a curb used as a miniretaining wall.

Fig. 8-36. Step face form in place. This form will not be removed after the pour. The nails sticking out of the inside of the board will be anchors to support the board against the concrete.

Fig. 8-37. Front view of Fig. 8-36.

horizontal, you will have to dig away the dirt and make room for additional steps. It gets complicated. Drawing sketches and plans can really help you. For this type of job, it is very important to have a good solid plan from which to work.

After you have decided on the pitch, length, width, and height of each step, you can prepare the side forms. You will have to use a form wider than the height of the step. This is because concrete will have to flow under each step face form and be solidly connected with the next step down the line. If there is to be a landing on top of the steps, it should be formed before the side (Fig. 8-39).

With the landing (if any) and the side forms set, you can begin forming the step face forms. Try to design the steps so that they are all equal in step surface (Fig. 8-40). After each step face form is placed, use the tape measure and the level to determine the location for the next one (Fig. 8-41).

The steps can be level, or they can have a slight downhill slope. As long as they do not slope into one another, you can adjust the pitch to form them at even lengths. The steps can slope so that each of them is 10 inches long, 18 inches long, 3 feet long, etc.

It is difficult to form and pour a set of rising steps, especially if the only experience you have is from the information in this book. It has taken professionals a long time to get the knack for step procedures, measurements, pitches, and concrete yardage estimates.

A good way to gain some hands-on experience is to contact the concrete dispatcher and ask him if he knows any contractors who are preparing to pour a set of steps. Go to the job site and ask the contractor if it would be all right to watch. Contractors usually will be glad to have you, and they can add a lot of information about concrete work while you're there.

Fig. 8-38. Steps must be formed so that they are evenly spaced and close to level.

Fig. 8-39. A small landing at the top of the steps.

Fig. 8-40. Try to design the steps so that each one will be the same length.

Fig. 8-41. Using a level to check steps.

Chapter 9

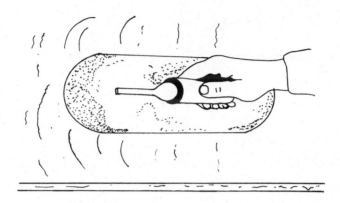

Setting Screeds

SCREEDING DETERMINES THE EVEN AND FLAT features of the slab. This phase demands attention. With the majority of walkways and patio slabs, the screed board can simply rest on the two side forms and be pulled (Fig. 9-1). If the slab you pour will be wider than 14 feet, however, you may have to install screed forms (Fig. 9-2).

A screed form is a 2 × 4 supported in the middle or along one edge of the slab, which will be a guide for the screed board. Raised so that the bottom of it will determine the top of the slab, the screed board will need an attached "ear" for support (Fig. 9-3). The ear is a wood stake nailed to the top of the screed board. It will rest on the top of the screed form, allowing the screed board to rest in the proper position and move concrete.

Any slab wider than 14 feet should have a screed form. Because of the long distance, lumber more than 14 feet will bow under the weight of the concrete. Even with three men on the screed board, the slight bow in the middle will cause a curve on the top of the slab. The amount of concrete having

to be moved by the screed board will make the task very difficult.

Besides slabs wider than 14 feet, screed forms should also be used when there are no side forms available for one end of the screed board. The rounded rock step offers no place for the screed board at that end (Fig. 9-4). The formers in this case have chosen to screed from the step toward the form. This is evidenced by the screed forms set up on each side of the step and resting on the form. The concrete will first be placed next to the step, working toward the form. Screeding will start at the step and work out. The small areas on each side of the step, not accessible by the screed, will have to be hand floated to properly place the concrete.

Another way to set up the screed form is to run it parallel with the end form (Fig. 9-5). The concrete will first be placed at one end of the screed forms, working down the length of them. The screed board will be supported by the ear resting on the screed form on one end and by the end form on the other. The screed forms should be pulled right

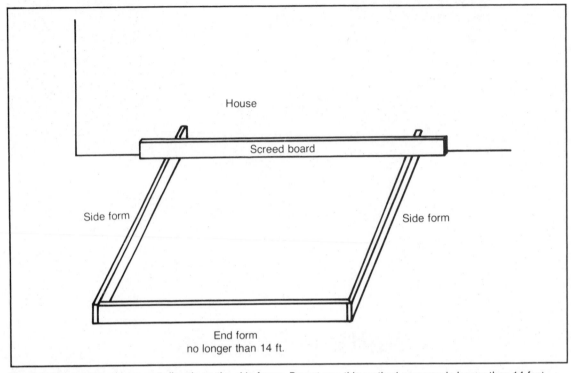

House

Screed board

Side form

Side form

End form
no longer than 14 ft.

Fig. 9-1. A screed board can rest directly on the side forms. Do not use this method on screeds longer than 14 feet.

Fig. 9-2. Screed form in position. The end of the board is resting on existing concrete for support and grade. A screed pin is supporting the middle of the board.

Fig. 9-3. Ear attached to a screed board resting on a screed form.

118

Fig. 9-4. Because no form is present next to the rounded step, screed forms were positioned at the sides of the step to screed the concrete away from the step.

Fig. 9-5. Screed form parallel with the end form. Screeding will go from left to right.

after that area of concrete has been screeded. The small holes left by the stakes can be filled in with a shovel, and tamping will flatten out any high spots.

SETTING SCREED FORMS

Setting screed forms is just like forming, except that the bottom of the screed form will determine the top of the slab rather than the top. The level will be your most valuable asset. Using it on top of the screed form will show you how high or low to place

the "free" end. One side of it can rest on a form. The other side must be supported by a stake. Before actually setting the level, you can nail the board to the stake. Just pound down the stake until the screed form is in place. You'll have to do this with the level on the screed form, so you can keep an eye on the slope.

CUSTOM SCREEDS

On unique pours, you may have to do some custom

screed forming (Fig. 9-6). Wide walkways with obstructions on one side call for a special method of screeding. The middle of the walkway will be screeded using screed forms. An ear attached to both ends of the screed board will support it on the screed forms. This method is time-consuming, but it will ensure a flat slab.

SCREED PINS

Special stakes just for screed forms are called *screed*

Fig. 9-6. A sample of custom screed forming. Because of the round projection, screeding will be easier coming away from it than going along it.

Fig. 9-7. A screed pin ready for the screed form. Adjustable arm is 1½ inches wide—just right for a 2 × 4.

pins (Fig. 9-7). A round stake is used to pound in the ground. The angled arm is adjusted on the stake by a small bolt. The distance between the stake and the upright part of the arm is 1½ inches—just right for a 2 × 4 form board. On a long stretch, a number of these screed pins can be placed in the middle of the slab area in a straight line. One screed form can simply be slid from one set of pins to the other as screeding continues. This will eliminate the need for additional screed form lumber.

Setting a screed form in the middle of the slab area is best done using a string and the screed pins. The string should be stretched from the tops of each side form. Be certain that it is stretched tight. A

Fig. 9-8. Pounding on a screed pin to lower the screed board to grade.

loose string will have a bow in the middle, and this will cause an error in the setting of the form. With the string as a guide, you can place the screed pins in a straight line. Adjust the arms to the proper height, and you are ready to lay in the screed form. If you didn't want to fool around with the arm adjustments, you could leave the pin high, lay in the 2 × 4 screed form, and then just pound the pin down until the bottom of the screed form matches the height of the string (Fig. 9-8).

The most common use for screed forms is for patio slabs poured against the house. If the dimensions were, say, 20 feet along the house and 10 feet away from it, you would screed the concrete using a 12-foot screed board going in the direction of the 20-foot length. Remember to screed the shortest side (Fig. 9-9). The 12-foot screed board will give

you enough board to reach the top of the end form and the screed form on the other side. The pour and screeding would start at the 10-foot side form furthest away from the concrete truck and work toward the other 10-foot side form 20 feet away. While tamping, the screed form could be removed and the stake holes repaired.

The best way to determine the need for screed forms is to visualize the screeding process during the planning stage. You then will have enough 2 × 4 lumber available and at the proper lengths.

SCREED BOARD EAR

When nailing the ear to the top of the screed board, allow the end of the ear to stick out past the end of the screed by at least 3 inches. The extra part extending out past the outer side of the screed form

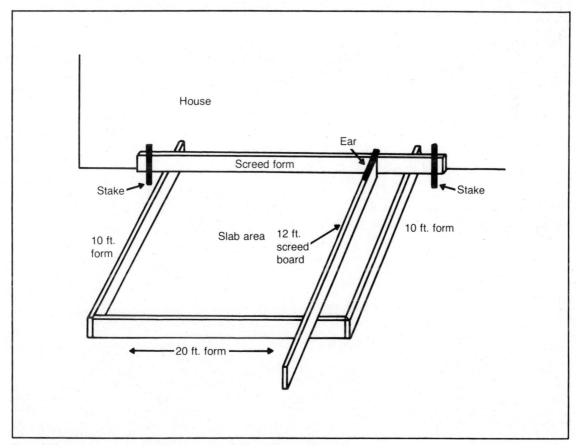

Fig. 9-9. Screed form set against a house to allow a 10-foot screed.

will let you seesaw the screed board across the concrete without it falling off the screed form (Fig. 9-3).

Try to visualize the entire job while you are drawing plans. Try to have every point covered throughout the job. Then you won't be in for any unexpected surprises during the pour.

Chapter 10

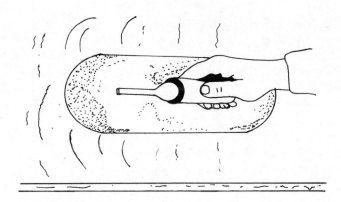

Forming Walkways

WALKWAYS ARE FORMED IN MUCH THE SAME way as larger slabs. The only difference is in the width. Walkways will also require expansion joints and seams. Because most walkways will stretch for quite a way, you should also be prepared to use a string as a straightedge guide.

Before starting to form, you should have an idea on the shape and width of the walk. A width of 3 feet, 1 inch will better accommodate the 3-foot-wide pouring and finishing tools.

An easy way to ensure straightness and place stakes is to use a forming block. The block is a 2×4 that is cut to precisely the same width as the walkway (Fig. 10-1). With one end of the block braced against the house, the form can be braced against the other end. By applying pressure to the stake, the form will be sandwiched between it and the block. The form and stake will remain rigid, and you can easily drive the stake straight into the ground.

Because each stake will be placed using the block, the form is in line with the wall of the house. The only problem you might encounter is if the wall

is not flat. Minor obstructions on the wall, as from extra bits of stucco, will cause the block to stick out too far in some places. If your wall is not smooth and flat, you will have to stretch a string and use it as your guide.

If you plan to have a planter between the walkway and the house, you will have to place a form on each side. For those jobs, cut a small block equal to the width of the planter. Place that inside form first. When it is in place, you can use the other block between the first form and the second.

SLOPE

As with other slabs, slope is important. Walkways can run at a level plane on one side, but they must slope on the other. For instance, the concrete can be level going from the back of the house toward the front. The slope must be from the house out. The concrete touching the wall of the house will actually sit a few fractions of an inch higher than the end 3 feet away from the wall.

On the other hand, if the walkway will butt

Fig. 10-1. Using a 2 × 4 block makes forming go faster and with more accuracy.

against a patio in the back and you want it to butt in the front at the driveway, the opposite could hold true. When the patio slab in the back sits higher than the driveway, the slope will naturally run from front to back. There will be no need to slope the concrete away from the house.

Slope is determined by how many quarter inches one end of the slab sits higher than the other. A widely used rule-of-thumb is to have the outdoor concrete pitch at ¼ inch per foot of run. For example, if a walkway was to slope from the backyard to the front and would be 20 feet long, the slope would be 5 inches. The end of the walkway at the backyard would rest 5 inches higher than the end in the front yard.

To adequately mark the slope, you'll need to use a chalk line. Determine the place where the top of the concrete will be at the backyard. Make a mark at that point on the wall. Find something on the wall that runs the length of it and is straight and level. You can use the mudsill lip on a stucco house, the edge of a board on a wood-sided house, or the bottom end of plywood sheets on a sheeted house.

Using your horizontal guide on the wall, follow it to the front. Measure down from it the same dimension as in the back (to the mark for the top of the concrete). Make a mark on the wall at that point. The mark on the wall in front should be level with the mark in back. After you have calculated the proper amount of inches for the drop, you can measure from the mark in the front down and make another mark. The second and lower mark will be the point for the top of the concrete at the front. Chalk a line from the mark in the back to the lower

Fig. 10-2. Bubble touching the second line on the level shows a good slope.

line in the front, and you will have the level marked for the run of the walkway. By placing the screed board on the chalk line at one end and resting the other end on the form, you can use the level on top of the screed board to set the height of the form.

To account for slope on the walkway form when no wall is available for a guide, use the level. Instead of measuring pitch and trying to get the form to an exact angle, I have used the lines on the level as a guide. Most levels have two lines at each end of the bubble. By angling the form so that the bubble just touches the second line on the high side, I have always had adequate slope (Fig. 10-2). When no other guides can be used and no matching points are in the plans, this method is quicker and easier

than any other I've found.

If your level does not have two lines at each end of the bubble, you can estimate the slope by having one-eighth to one-sixth of the bubble past the first line. You must know how far the bubble goes past the line at the start. Then you will have to make sure all other readings are taken with the same degree of bubble extension.

MATCHING

Matching the walkway from a patio in back to the driveway in front is the best way to run it. In most cases, the backyard is generally landscaped higher than the front. Therefore, the slope for the walkway will be a natural. In those cases where they are

Fig. 10-3. Matching the walkway with the concrete in the front of the house.

Fig. 10-4. Assorted screed forms set up on a difficult job.

even, you'll have to slope the walk away from the house (Fig. 10-3).

A string attached to stakes at each matching point and set at slab level will give you a guide for form height. The screed board, with the level tool on top of it, will have to rest on the staked form with the free end raised or lowered to make the height point. Starting out at one end of the walkway, use the screed and level to make a mark on the wall. Do the same thing at the opposite end. Chalk a line

between the two marks. That line will designate the high point of the slab. The lines can be used as a guide for setting screed forms, or it can be the sole guide when pouring the concrete.

WALKWAY SCREEDS

Chapter 9 details the setting of screed (Fig. 10-4). For walkways, there is another way to screed the concrete.

Chalk lines can be used as screeding guides.

One line will mark the top of the concrete. A second chalk line can be made exactly 3½ inches above the first. The second line will be a guide for the *top* of the screed board.

Because wet concrete will cover chalk lines, you might lose your guide during the pour. With the second line adjusted for the height of a 2 × 4 screed board, it can be used in lieu of the first. The chances of the second line being covered with concrete are low.

After the concrete has begun to flow into the forms, use a hand float to push and flatten the mud to the line. Attempt to hand screed the length of the hand float about 12 inches from the wall. This 12-inch section of screeded concrete will serve as an additional screed guide along with the upper chalk line. The term used for this type of job is *wet screeding*.

LAYOUT

A walkway should be laid out just like a slab. Placing the forms in their approximate locations will give

Fig. 10-5. String will line up the walkway past the curved turn.

you the chance to make the best of form length placement. Use the forms to their greatest advantage. Do not cut long forms unless absolutely necessary. Many forms can be put to very good use after a concrete job. Some can be used to build a patio cover; others can be used to build shelves or repair fences. Don't waste the form lumber.

Some finishers like to string the job before they begin forming. They place stakes in strategic locations and attach string to them. The string will give them a visual guide as to where the forms will go and what the job will look like when formed.

String can be useful in many ways. Besides use as a height guide and straightedge for forms, it can also help to line up forms that are separated by open areas (Fig. 10-5). When stretched tight, the string will be straight both in the vertical and horizontal planes. The important factors in string use are the locations of the stakes holding it in place and its height when tied to the stakes.

EXPANSION JOINTS

Walkways should have an expansion joint every 10

Fig. 10-6. Stringers used as design and expansion joints.

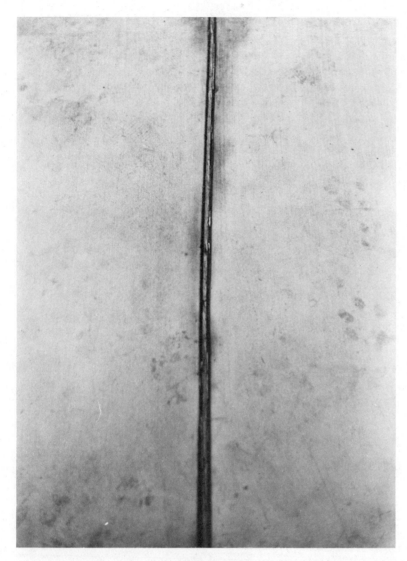

Fig. 10-7. Benderboard strip used as an expansion joint.

to 12 feet. They should be divided by seams every 3 to 4 feet. The three basic kinds of expansion joints are felt, stringer, and benderboard.

Felt is supported inside the forms by a 2 × 4 brace. Essentially, you will have to place a form inside the slab area. The stakes should be placed on the side of the form that will receive the concrete last. The stakes will brace the form against the concrete first laid. That concrete will support the form after the stakes are removed. The concrete should be placed against the felt first. Additional concrete should then be laid behind the form, sur- rounding the stakes and form. When about 1 foot of concrete is down, the stakes can be removed. Gently remove the 2 × 4 form while someone else fills in the empty spot with concrete from a shovel. The procedure should be done in such a way as to not disturb the felt. When you are done, the felt will be securely in place and straight.

Stringers

Because felt is so hard to work with, you might consider the other two options. The first is string- ers, which perform the task of an expansion joint and also look nice (Fig. 10-6). Placed inside the forms, the ends can be nailed to the side forms. For most 3-foot-wide walkways, that is sufficient. If no form is available on one side, however, stakes will have to be used. The stakes can be pulled after the concrete has been placed around it.

When using 2 × 4s for stringers, you should always place nails along the sides of the wood. Driven halfway into the sides of the stringer, the extending part will be encased in concrete. This action will have the nails working as anchors for the stringer. With them, you will never have to worry about the stringer "floating" up out of the concrete.

Because most wood will rot after being subjected to weather for prolonged periods, you should not use ordinary wood as stringers. I suggest using either redwood or pressure-treated Douglas fir. Redwood is pretty and can be stained with a clear finish to preserve the grain features and color. If you plan on staining the stringers with a dark color, you may find it cheaper to go with pressure-treated fir. The pressure-treated wood will be marked with many tiny slits. Barely noticeable, they will not cause an eyesore. Both types of wood are commonly available at local lumberyards.

Benderboard Expansion Joints

The third method of making an expansion joint is with thin strips of benderboard (Fig. 10-7). By splitting a 4-inch piece down the center, a 2-inch-high board can be easily inserted into wet concrete. After the concrete has been tamped and bull floated, a section can be wriggled into place. Lengths longer than 3 feet are difficult to place. Therefore, this method is really only adequate for 3-foot walkways (Fig. 10-8).

Nails placed on the side of a form can be used as guides for benderboard joint placement. Your biggest concern will be to get them in straight. Use a hammer to gently persuade the strip into place. Afterward, a hand float can be used to smooth the concrete.

If a benderboard expansion joint sticks up after the concrete has cured, it can be cut flush with the surface. You can either use a razor knife to cut it or

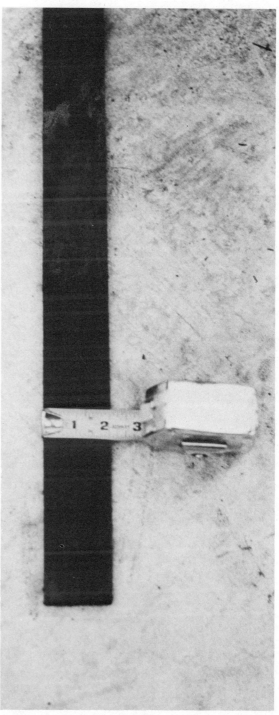

Fig.10-8. A 2-inch strip of benderboard is ready to be inserted into the concrete.

133

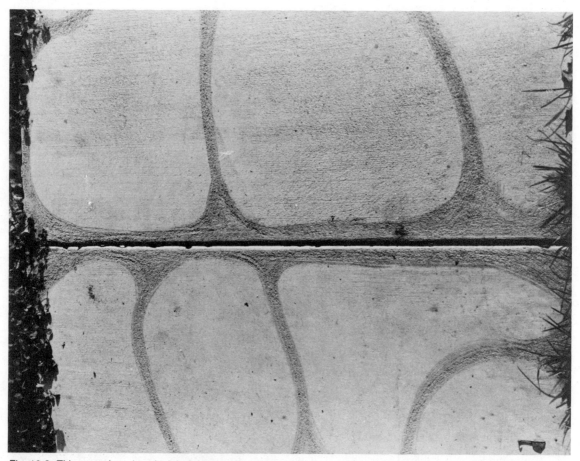

Fig. 10-9. This seam has done its job. It has prevented the crack from surfacing somewhere else on the walkway.

run a shovel blade over the top of the concrete to slice off the excess.

Seams will be placed in the walkway between the expansion joints. Their purpose is to control cracks (Fig. 10-9). Because the seams do not completely separate the concrete, they are not supposed to prevent cracks. As a control joint, they will allow the concrete to crack inside the seam and make it relatively unnoticeable. The crack inside the seam in Fig. 10-9 does not detract from the appearance of the walk, whereas a crack along the etched surface would be an eyesore.

CUSTOM WALKWAYS

Walkways do not have to be plain looking sidewalks. You can develop a custom walk by doing

some work with the forms.

Large steps don't look like walkways, even though they have the same use and are formed in much the same way (Fig. 10-10). By adding a form or two inside the walkway forms, you can separate sections of the walkway and fill them in later with rock, decorative bark, or grass.

Straight walkways with curves in them look much nicer with some landscapes than right-angled corners (Fig. 10-11). Free-formed walks can help to outline a planter (Fig. 10-12).

Landscapes utilizing many trees and large shrubs make little room for straight walkways. Using a combination of both straight and curved features around trees will give your yard a parklike appearance (Fig. 10-13). Dressing the walkway with a custom finish using stringers and a rock salt, exposed, or etched finish will complete your custom walk.

Unique angles formed for a walkway also create a custom job (Fig. 10-14). Forming unusual angles takes a little more time and work. The ends of some forms will have to be cut at angles equal to the angle you are producing for the concrete. By laying the form to be cut on top of the straight form, you can mark the angle of the cut with a pencil. The pencil mark will have to be made on the bottom edge of the form because that is where the corner of the straight form will be. After getting the mark to show the angle, use the square to make a straight line down the side of the form. With the angle mark as the guide for the angle cut, the straight line will guide you in making the cut perpendicular with the board (Fig. 10-15).

Finally, slight angles on walkways break up the solid concrete look (Fig. 10-16). Narrow walkways can be spruced up with angles, 1 × 4 stringers, and an exposed aggregate finish. Forming custom walkways will take more time and more money if stringers are used. The time and money will be a small sacrifice for the amount of pleasure you will get from the job.

Fig. 10-10. A unique step design with grass strips separating each step.

Fig. 10-11. Combination straight walkway with a curved twist in it to meet the front steps.

Fig. 10-12. Free-form walkway serving as a place to walk and as a border for the planter.

Fig. 10-13. Use of a curve around trees with an otherwise straight walk.

Fig. 10-14. Unique angles for a custom front entry and walkway.

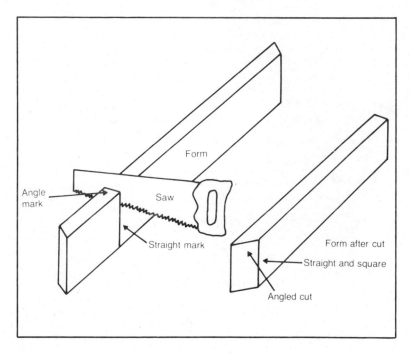

Form

Saw

Angle mark

Straight mark

Form after cut

Straight and square

Angled cut

Fig. 10-15. When the top or bottom of a form is marked at an angle, make a straight line down the side as a guide for a straight cut along the 3½-inch side.

Fig. 10-16. Narrow walkway with 1 × 4 stringers and a slight angle to better accommodate the landscape.

Chapter 11

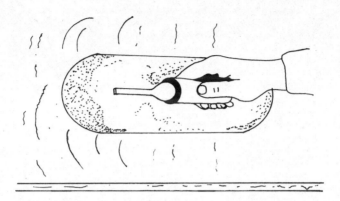

Forming Curves

A YARD LANDSCAPED WITH MANY CIRCULAR planters may be finished with curved designs in the concrete. Some curves only need a piece of benderboard added inside the forms. Others, especially the free-form designs, will be completely formed with it.

ROUNDED CORNERS

Making a curved corner rather than a 90-degree one will be functional and unique (Fig. 11-1). The only material needed to form the corner will be benderboard and ¾-inch roofing nails. If benderboard will not line the forms completely, you will also need a rasp to smooth the protruding lip.

To effect the curved corner on a walkway, first form the walk as if it was going to be totally straight. When all the forms and stakes are in place, insert the 8-to-10-foot piece of benderboard into the corner. With both sides resting against the 2 × 4 forms, push the center of the benderboard into position. When the correct arc has been made, place a stake in front of the arc to keep the benderboard in

place (Fig. 11-2). Move to one end of the benderboard and nail it in place.

To secure benderboard to the 2 × 4 forms, use several ¾-inch roofing nails. The thinness of the wood allows it to split easily and prevents large duplex nails from being effective. The large head on the roofing nail will keep the benderboard in place, and the short shaft will not split the wood quite as quickly. Place the nails at different locations and heights on the benderboard. Putting two nails along the same horizontal line will cause a split.

Nails will have to secure the benderboard from the extreme end all the way down to where the benderboard starts to separate from the 2 × 4 form and begin the curve. Generally, this will take about six to eight nails.

After the first side is nailed, with the top matching the top of the 2 × 4, proceed to the other end. You can still move the end of the benderboard and change the arc of the curve. When the arc you want is established, nail that end of the benderboard to the 2 × 4 form. With both ends nailed, your curve is established.

Fig. 11-1. A curved corner for a walkway.

Fig. 11-2. Stake holding the arc in place while installing benderboard.

Redwood benderboard is very brittle when dry. Wet it before using it. When extremely tight arcs are desired, you must soak the benderboard in water (Fig. 11-3). Allowed to remain dry, the benderboard will crack and break in half. Instead of wasting money by breaking forms, always wet down benderboard before using it.

Forming this type of curve will result in a lip protrusion caused by the ends of the benderboard. To solve that problem, you can line the entire 2 × 4 formed area with benderboard, or you can remove the lip.

A perfect tool for removing the lip is a rasp (Fig. 11-4). Working the tool back and forth will wear off the lip and create an angle at the butt of the benderboard The angled surface will permit the concrete edging tool to glide smoothly from the 2×4 to the benderboard. This action will allow you to place a clean edge on the concrete.

MATCHING CORNERS

Some walkways and most patio slabs have two corners. Curving one and not the other would look funny. One curve at a tighter arc than the other

Fig. 11-3. Dry benderboard will split when you try to bend it. Soaking it in water will replace the elasticity.

would also look awkward.

You'll need to take some accurate measurements to match the corners. Measure the distance from the end of the benderboard to the point where it begins to separate from the 2 × 4 form. Take measurements from both ends of the benderboard. Use those dimensions to make the identical readings on the corner to be curved. Mark the 2 × 4 form with a pencil at the points where the ends of the benderboard should be and where they will begin to separate from it.

Go back to the curved corner. Measure the distance from the actual corner of the 2 × 4 forms to the center of the arc in the benderboard (Fig. 11-5). Make a note of that figure for future use.

At the uncurved corner, place the benderboard into the corner using a stake in the center as you did with the first corner. Place the stake the same distance away from the intersection of the 2 × 4 forms as the first one. Using the pencil marks on the 2 × 4 form, nail one end of the benderboard in place.

Go back to the arc and readjust the benderboard until it is the same distance away from the 2 × 4 intersection as the first curve. Place the stake for support. The end of the benderboard not nailed should be very close to its pencil marks. If the end matches up and the arc is correct, you can nail that end to its 2 × 4 form. You will have to rasp the ends of this piece just like the first. The corners are completed with two identical arcs.

Forming curved corners inside formed 2 × 4 areas is not hard. Keep the height the same as the 2 × 4, measure and keep the arcs even, and rasp the protruding lips. Later, I'll cover how to brace the corner and make it ready for concrete.

FREE-FORM SLAB

Setting up a free-form slab takes imagination and

Fig. 11-4. Use a rasp to remove the lip caused by the benderboard.

Fig. 11-5. To measure the arc in the benderboard, lay the tape in the corner of the 2 × 4 forms and allow the end to touch the benderboard.

patience. When making tight curves, you'll have to use benderboard. Its thin design makes it difficult to steady, nail, and keep in place. Many stakes have to be used with it, both outside and inside the formed area. Using 1 × 4 material when forming slight curves is much easier to work with (Fig. 11-6). Because there is more area on the board, it is more rigid. This feature makes it less brittle and easier to nail. Although bending it into position will take a little muscle, the overall advantages are worth the effort (Fig. 11-7).

When free forming, have the area graded close to the grade desired for the concrete. This will make placing the forms easier and will not require that you move them once they are in position. With the grade close, you should place a few stakes along the point where the form will lay. Place them in such a way as to almost outline the job. Then go ahead and start putting in the benderboard.

Use stakes on both sides of the benderboard form to sandwich it in place and keep it supported. When the design has been determined, you can start checking for grade height. Use the screed board with the level on top of it to help locate the proper form height. The opposite end of the screed must rest at its proper point to keep the form end accurate. Check the level continually to assure accuracy.

The base may have to be graded at this point to allow the form to be pushed down. If that is the case, simply pull the benderboard up a couple of inches. The sandwich action of the stakes should keep it in relative position. Use the shovel to establish grade and then put the form back down. When the height for the entire form is set, you can begin nailing the form to the stake.

Fig. 11-6. One-by-four is easier to work with on long sweeping curves.

Fig. 11-7. Free-form side on a patio slab.

Fig. 11-8. One-by-four curved forms can be staked just like 2 × 4 lumber. Watch out for nails coming through when nailing stakes.

Nailing benderboard to the stakes is different than with 2 × 4 lumber. You will nail the benderboard with the ¾-inch roofing nails from inside the forms. There is not enough wood on benderboard strips to hold a regular nail. By nailing from the inside with the large-headed roofing nail, there will be no problems. Stripping the forms will entail taking the stakes a little differently. First, you'll

have to pull the stake away from the form. While it is at an angle and away from the nail, you'll have to wriggle it out of the ground. That's why you should use shorter stakes with benderboard. Added support for the forms can be achieved in several ways. I will discuss them later.

Securing 1 × 4 curved forms is easier. Because the wood is thicker, you can use duplex nails

Fig. 11-9. To double benderboard for strength, lay two pieces together and put nails through them. Bend over the nails.

but you should use 6d nails. The shorter shaft will not go through the wood and into the concrete area. When bracing your foot against the form while driving the nail in from the stake side, be careful not to put the nail into your foot. Position yourself so that the nail will be pointed to the side of your shoe (Fig. 11-8).

REINFORCING UNSUPPORTED BENDERBOARD

Forming curved steps presents some special prob-lems. First, you have to get the arc even. Second, the forms must be sturdy. Third, the stakes must do an adequate job.

Because benderboard is so thin, doubling it up with two pieces together will give greater support (Fig. 11-9). About the only way to secure the two pieces is by putting many nails through them. The heads of the nails will secure the top piece. Bending the shafts over will secure the second (Fig. 11-10). With one accurate length of double thick bender-

Fig. 11-10. The bent nail will hold one side while the head of the nail holds the other.

Fig. 11-11. Place a stake at each end you want the form. Move the form and stakes around until you establish the arc wanted.

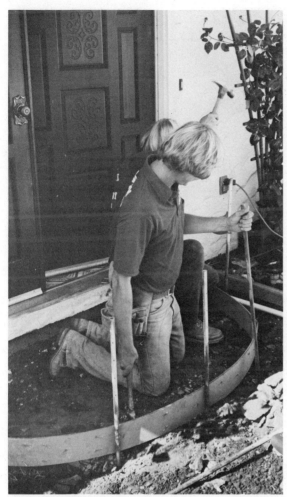

Fig. 11-12. When the arc is made, use stakes to hold the form in place.

place. Once the concrete is against the forms, there is no need for a stake to support the same thing.

The professional concrete formers shown here are using metal stakes to support benderboard. They use long nails bent over on the inside of the form (Fig. 11-14). To me, this is wasted time and work. Using wood stakes and ¾-inch roofing nails is quicker and easier. You don't need two men to nail, and the stakes and forms are easier to pull.

When the stakes are in place and the arc is supported, you can concentrate on the height. On a step that is part of a continuing walkway or patio, you can stretch a string from the top of one side form to the other. The string can be used as a guide for the bottom on the step face form. In conjunction with the string, a level can show you the proper slope (Fig. 11-15).

The far end of the step can rest on the string. The ends against the house can be moved up or down to create the proper slope. When the step is formed, check the entire length of the benderboard for weak points. Places where the form can easily bend out of shape should be reinforced with more stakes.

When the step is totally secured, raise the base depth to its proper height (Fig. 11-16). The center of the step should be 4 inches. On the sides, an open area that goes down to the rest of the slab should be allowed. This open space will let the concrete flow together and be joined to add strength.

INSIDE CURVED BORDERS

Extra custom work is achieved by installing a benderboard border inside the slab area. The result will appear like a curved slab surrounded by a border of concrete. The outside border often is colored. To produce this kind of work, you must be on your toes and know exactly what you are doing (Fig. 11-17).

The first thing that must be done is to set the forms for the outside edge. That would be the straight 2 × 4 forms (Fig. 11-17). After the job site is graded, you can begin with the benderboard.

The curved ends should be done first. Establish the arc by using stakes as supports. When the

board, you can begin forming.

Establishing the arc takes a little time. Use stakes at each end of the form as braces. The stakes can be moved closer or further apart, creating tighter or wider patterns (Fig. 11-11). Once the arc is established, you can start placing stakes. Don't worry about the height yet. You have to get stakes in place first to support the arc (Fig. 11-12). Because this step (as seen in the illustrations) is going to be surrounded by concrete, the stakes will be removed after the concrete has started to set up (Fig. 11-13). The stakes supporting the inside of the forms will be pulled after the concrete is in

Fig. 11-13. Support stakes will have to come out after the concrete is poured. Use a trowel or putty knife to fill in the holes.

Fig. 11-14. Using metal stakes with benderboard is harder than using wood stakes and ¾-inch roofing nails. In this case, two formers must work together to bend the long nails over the benderboard.

correct arc is made, nail the benderboard to the stake. Use the screed board to show the correct height. Pound down on the stakes as necessary to get the form at its right height (Fig. 11-18). When securing this type of benderboard forming, you can pound the stakes down a good 1 inch to 1½ inch below the top of the benderboard form. By doing this, you can leave the stakes in the ground after the pour. If the ground is too hard for driving the stakes down far enough, use the power saw to cut the tops of the stakes.

Placing benderboard in a straight line can be made easier with the use of blocks (Fig. 11-19).

When placing the stakes, brace your foot against the stake. Sandwich the form between it and the block. The block will be supported by the 2 × 4 form (Fig. 11-20). The blocks should be left in place during the pour. After the concrete is in place, the blocks can be removed and the holes filled with concrete from a shovel. Instructions on pouring the concrete for this type of job are in Chapter 14.

BRACING BENDERBOARD

Because of the flimsiness of benderboard, you must be sure it is securely braced. There are three basic ways to do this.

153

Fig. 11-15. Use a level and a string to guide forms to their right height.

Another way to brace benderboard is by staking 2 × 4 pieces behind it (Fig. 11-23). This works great if the benderboard is not formed in tight arcs. This method can take a lot of time, and you will have to have enough stakes and short pieces of 2 × 4.

Probably the cheapest, quickest, and easiest way to brace benderboard is with dirt (Fig. 11-24). Damp dirt makes a great support. The dirt cannot be used in lieu of stakes, but it makes a good partner to them. By placing about 6 to 8 inches of dirt—the height of the form—behind the benderboard, you can be assured of a firmly supported form.

First, you can use many stakes placed every 1½ feet (Fig. 11-21). This will be especially true if the benderboard makes an actual form and will not be supported by the concrete during the pour.

When placing long expansion joints using benderboard, you will still have to use many stakes. It usually is easier if you sink the stakes down 1 to 1½ inches below the top of the form. Then you can cover them with concrete and just leave them. No messy pulling is needed (Fig. 11-22). When the concrete is poured, you'll have to put an inch down against one side, then an inch against the other. Evenly placing the concrete slowly against each

Fig. 11-16. On steps, fill in the middle of the area so that it is no deeper than 4 inches. The sides of the step must be open to the area underneath for support.

Fig. 11-17. Custom concrete job ready to pour. The redwood benderboard inside the forms will form a border of red concrete around an exposed aggregate finish.

Fig. 11-18. Screed board used to help set height of benderboard.

Fig. 11-19. To form straight bender-board, use blocks as supports.

Fig. 11-20. When pounding stakes against blocks, use your foot to put pressure on the stake, which will sandwich the benderboard and keep it in place.

Fig. 11-21. Using benderboard forms requires stakes about every 1 to 1½ feet.

Fig. 11-22. To completely secure benderboard, place stakes in the ground 1 inch to 1½ inches past the top of the form. Nail the stake to the benderboard.

Fig. 11-23. Two-by-four boards can be staked behind benderboard for support.

Fig. 11-24. The cheapest and quickest way to brace benderboard is with dirt.

Chapter 12

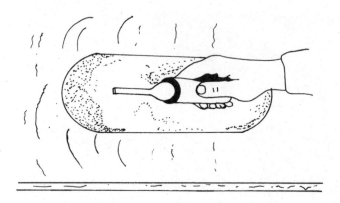

Redwood Stringers

STRINGERS IN CONCRETE CREATE A CUSTOM and eye-appealing slab (Fig. 12-1). They can be used as mere decorations or can play a useful role as an expansion joint (Fig. 12-2). The stringers you use should be of redwood or pressure-treated Douglas fir. After they are inserted and securely planted in the concrete, you will never be able to replace them if they should rot. Redwood and pressure-treated lumber will not rot, so you don't have to worry about replacement.

DESIGN

Stringer designs are almost limitless. Any pattern imaginable can be instituted (Fig. 12-3). Any concrete finish can also be applied along with the stringers.

One redwood stringer can be used to section a slab in two equal halves. You can use more to design a slab in more equal parts. You can even use enough stringers to form a slab into 34 separate parts, leaving a few of the sections open for use as planters. There is no limit to the patterns. The boards do not have to run perpendicular with each other or at right angles with anything. Stringers can form diamonds, octagons, triangles, and different sized rectangles.

Before actual forming, draw several sketches. Have a very good idea of what you are going to design before you actually start cutting and fitting stringers. Redwood is expensive wood. A miscalculated cut can ruin the appearance of a redwood stringer. It is far better to have a preplanned set of blueprints to go by, ensuring accurate stringer placement, cuts, and purchase. You might loosely lay out the stringers in a designated pattern before you start nailing. This will show you what they will eventually look like.

STARTING

The simplest place to start is at the point using the longest stringer. For example, if you have decided to use a pattern that divides the slab into four equal parts, lay the lengthwise stringer first. Measure along the side forms until you get the center point

Fig. 12-1. Stringers break up the solid concrete look and add much to a slab.

established. Make a mark with the pencil at the center on top of the side form. Do the same thing to the opposite form. Measure from the same reference point. If you measure out from the house on one side, do the same for the other. This will ensure an equal distance for both the forms.

After both side forms have been marked at their center points, lay the stringer inside the forms. Position each end of the stringer at the marks on the side forms. Check the entire length of the stringer to be sure the grade is proper. No part of the board should be higher than the tops of the side forms. You may have to use a string attached to the side forms to confirm the grade.

If the grade is all right, nail one end of the stringer to the side form. Be certain that the middle of the stringer matches the pencil mark on the side form. Nail from the outside of the form (Fig. 12-4). After the first side is nailed, nail the other side with duplex nails. Unless the outside forms are to be left

in place after the pour, you will want to pull the nails when ready to strip the forms.

When you have both ends of the stringer nailed with two nails in each end, you can use the string as a straightedge guide. Starting at one end of the stringer, adjust it until it is perfectly straight. Place stakes every 4 feet or so to secure the board.

On short runs of 12 feet and less, you may put stakes on only one side of the stringer. That side should be the one that will get its concrete last. The stakes will support the board while the concrete is being poured on the opposite side. When concrete is laid on the staked side, the previous concrete poured on the other side will be the supporting brace. This concept is to be used when you are placing stakes and not nailing them. On short runs where the grade is solid, you won't have to nail the stakes in the middle. The stringer will be supported on each end and will only need bracing because of the concrete's weight.

Fig. 12-2. Stringer used as an expansion joint in patio slab.

Fig. 12-3. A custom stringer pattern. The white material on the slab is rock salt.

nails along both sides of the first stringer. Using 8d nails or larger, drive a nail partially into the sides of the stringer every 2 to 3 feet. Allow at least half of the nail to stick out (Fig. 12-6).

These nails will be encased in the concrete and will hold the stringer in position vertically long after the concrete has set up. It doesn't happen often, but periodically stringers have been known to float up out of the concrete. A small rise in the stringer can cause a dangerous hazard. People walking by the "risen" stringer can trip on it, causing injury. Anchor nails prevent floating stringers and will keep your slab safe. All stringers must have

If you think additional support is needed because of grade, design, or the need to roll wheelbarrows over the stringers, you can insert stakes and nail them, too. To do this, you will have to sink the stakes 1½ inches below the top of the stringer (Fig. 12-5). This will allow concrete to surround the stake and leave enough mud over it to be finished. When staking, try to remember that you will be placing another stringer intersecting the first. Don't put a stake where the next stringer will be placed.

When the first stringer is placed and formed, measure to the middle of it and mark it with a pencil. After the center is marked, use the small hand square to make a straight line across the top of the stringer ¾ inch on each side of the center mark. The distance between the two ¾-inch lines should be 1½ inches—the actual width of the 2 × 4 stringer. In a few minutes, you will be making a 1¾-inch cut down both of those lines.

ANCHORS

Before moving to the second stringer, place anchor

Fig. 12-4. Securing a stringer by nailing it to the outside form. The nail should be driven from outside the form, so it and the form can be pulled later.

Fig. 12-5. Pounding stakes down about 1½ inches below the top of the stringer. The stakes will be nailed and left in the concrete.

anchor nails attached—even ones that are smaller than 2 × 4s. The only exception would be bender-board. The thin width of benderboard allows for the weight of the concrete to keep it in place. Because it is so skinny, it doesn't shrink as much as a 2 × 4 and won't float.

SECOND STRINGER

The first thing to do with the second stringer is to measure it. Lumber doesn't always measure the length it is suppose to be. A 10-foot board might measure 10 feet, 2 inches. Measure the distance that the second stringer is going to cover. Measure the stringer for the same distance and cut off any

extra. Let's make it 10 feet for this example.

At exactly 5 feet, and measure the board from both ends to be certain that the 5-foot mark is in the middle, mark the top of the stringer with a pencil. Just like on the first one, make a straight line ¾ inch on each side of the middle mark. After this, use the hand square to make a 1¾-inch line down one side of the board. The 1¾-inch line will be a continuation of the ¾-inch line on the top of the stringer. Make a 1¾-inch line for each ¾-inch line.

These lines will be guides for notching (Fig. 12-7). The ¾-inch lines on the top of the stringer will allow for a 1½-inch-wide notch. The 1¾-inch line will be a guide for the depth of the cut. After all the lines have been made, cut the wood precisely on the lines.

Fig. 12-6. After the stringer is placed and secured, put a few nails halfway in the sides of the stringers. After the concrete has set up, the nails will be anchors for the stringer.

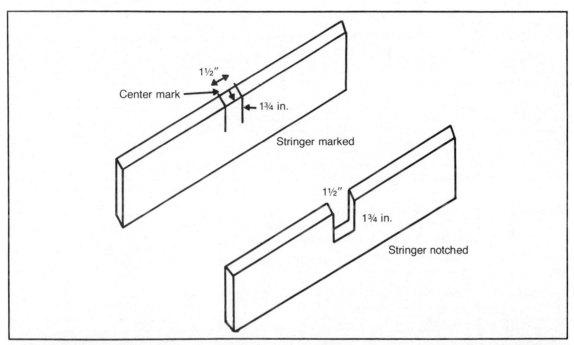

Fig. 12-7. Top board shows how to mark for a notch. The bottom board shows the notch—1½ inches wide and 1¾ inches long.

Fig. 12-8. End of the second form is nailed to the end form.

When the cuts have been made extending halfway down the side of the stringer, lay the board down. Use a hammer to firmly smack the chunk of wood between the cuts. This action should knock the chunk out of the board. Use a chisel to clean up the notch.

On the stringer already placed, make the same cuts on the lines you had previously made. Use a hammer to knock out the notch and a chisel to clean up the burrs. When you are done, you should have two identical notches. They should both be 1½ inches wide and 1¾ inches deep. These dimensions will allow both of the stringers to be fitted together.

Pick up the free stringer and place its notch over the notch of the staked stringer. By pushing down on it, you should be able to position it completely down with its top flush with the other. You

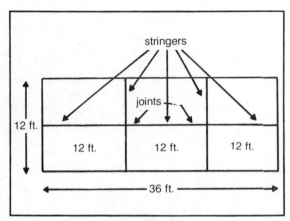

Fig. 12-10. By using three 12-foot stringers the length of the slab, the ends of all stringers butt against the center stringers. There is no free floating joint in the center of the slab, as there would be using two 18-foot stringers.

may have to use a hammer because it will be a tight fit. If the fit is too tight, you may have to widen the cut by just a bit. Use a rasp for that action.

After the second form has been fitted to the first, you can stake it and nail its corner to the end form, providing the other end will be against the house (Fig. 12-8). Staking procedures will be the same as with the first stringer (Fig. 12-9).

Although notching takes more time, it makes for a better job than just butting the ends of two stringers to the sides of the first. Notching guarantees that the joint will be square and the side runners will not get moved during the pour. It is also easier to notch the boards than to accurately measure and stake two boards in a straight line.

Notching should be done with all intersecting stringers. Whenever possible, use one board to make the run. It is difficult to place two stringers in a perfectly straight line and keep them in that position during the pour, especially if they are not staked as well as outside forms. Notching adds support, maintains alignment, and makes for a custom job.

UNIQUE STRINGER PATTERNS

When buying the material for your stringers, make sure that at least one 1½-inch side is perfect. Knots, cracks, and mars on the side that will be visible after the pour will ruin the appearance of the

Fig. 12-9. Stringer support stakes will be pulled after the concrete has been placed against both sides.

Fig. 12-11. Planters made out of stringer sections.

stringer. Straight boards are essential.

If you are going to construct a unique design, try to buy stringers that will fit in one piece. Obviously, if your slab will be 36 feet long, you won't have to buy a 36-foot board. Depending on the design, you might not want to use two 18-foot boards.

Let's imagine a design that will split a 12-foot-by-36-foot slab into six separate sections (Fig. 12-10). The two stringers that run from the house to the end form will be 12 feet long. If you used just two 18-foot stringers, they would butt together in the middle of the concrete. If you used three 12-foot stringers, each end would butt next to a stringer.

There would be no free floating joint at all.

This logic applies to all form purchases. Determine beforehand what lengths you will need. There is no reason you should have to buy more lumber than will be used for the job. Wasted money offsets the money saved by doing it yourself.

When designing patterns for stringers, you might want to allow some sections free of concrete to be used as planters (Fig. 12-11). When planning for this, don't forget to calculate only those sections when estimating for concrete yardage. Measure each section separately that will be filled with concrete. Add all of them together to get the final figure.

Fig. 12-12. With long stringers, you will definitely need a walking edger with extensions.

PLACING THE CONCRETE

Special problems sometimes arise when pouring concrete around stringer slabs. Wheeling over the 2 × 4 stringers is an obstruction that will have to be accounted for. Heavy-duty ramps must be made over each stringer that will be in the way. Extra staking on each side of the ramp will help keep the stringer in place. If at all possible, avoid going over them. On medium to large pours 3 yards and up, you might consider having the job pumped if the slab area is interspersed with many stringers.

Going over one or two stringers with a few

wheelbarrows of concrete poses no real problem. Having to roll over five stringers when wheeling 4 to 5 yards of concrete is a real hassle. Pouring directly from the truck poses no problem at all, especially if the chutes will reach all of the slab area. For your own slab, use common sense and try to visualize the job. Placing concrete is a demanding job; make it as easy as possible.

EDGING AND CLEANING STRINGERS

All stringers should be edged. Just like along the outside forms, the concrete edging tool must be run along all sides of all stringers. This maneuver will clean the edges and dress up the job. For the most part, you should have no problems. Try to use a walking edger with extension adaptability (Fig. 12-12). This tool will greatly aid in edging stringers. Edging them with hand edgers is very difficult. You will have to tiptoe along the stringer and bend down to effect the edge. This will result in footprints in the wet concrete that will need repairing.

If your job makes use of more than three stringers, you should have enough helpers so that

Fig. 12-13. Applying a protective sealer on the top of the stringer.

168

two of them can be responsible for all the edges. This is an important task that will take some time. Having helpers ready and available to take care of all the edges will free you to concentrate on fresno work and other tasks.

On the job I did with 34 separate sections, I had one worker do nothing but put on edges. By the time he got done edging the entire slab once, the concrete was just about ready for another. As the concrete sets up, edging and finishing will have to be done more often. Having that worker take care of the edges allowed me to continue with the rest of

the work, and the slab turned out fine. Trying to do it by myself would have exhausted me, and the finish and edges would not have turned out as nice.

PROTECTING STRINGER TOPS

The tops of the stringers will get concrete on them. Some finishers have used heavy-duty duct tape on the tops to protect them from concrete buildup. Others have cleaned the tops with a putty knife when they were out on the slab finishing. Other finishers like to put a coat of protective sealer on the stringer before the pour (Fig. 12-13).

Fig. 12-14. Follow directions on the label of any sealer you use. Also, note the anchor nails sticking out of the stringer.

I have had bad luck using tape. The water in the concrete can loosen the tape. The loose tape falls off the stringer and into the concrete. This is really a hassle during screeding when 2 × 4 screed boards are rubbed across the tops of the tape. It is hard to get the tape out of the concrete once you have started to bull float and fresno. Tape may work fine on small jobs, but I would not use it on large ones.

Cleaning the tops of the stringers during the pour is extra work, but it's not that bad. While tamping, you can bend down once in a while to clean them with a hand trowel. Later, while finishing, use a putty knife to really clean them. You will have to use the hand edger next to the stringer anyway, so it doesn't really make the work that much more difficult. After the concrete has fully cured (about two weeks), you can use a wire brush to completely clean the tops and make them like new.

Special agents used to seal stringers are available at concrete plants, lumberyards, and hardware stores. Some finishers have successfully used motor oil and diesel fuel to prevent concrete from sticking to forms. On a nice piece of redwood, you might want to use an agent that will allow stain to adhere later. Ask your local concrete dispatcher for advise. Climate of your area will determine what agent you should use (Fig. 12-14).

Stringers can make screeding a breeze. Stringers that are securely braced and supported make wonderful screed forms. You might incorporate that idea into your slab. If the job you want to do is going to take a screed form down the middle of it to guide the screed board, you might consider installing a stringer instead. The stringer will serve as a screed form, an expansion joint, and even add a touch of flair to your work.

Chapter 13

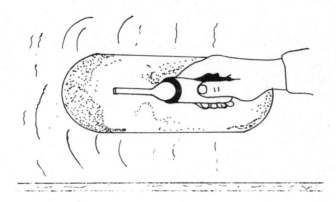

Other Considerations Before Pouring

MANY TIMES PEOPLE HAVE POURED CONCRETE and later regretted not placing drains under it, or they wished a pipe sleeve would have been installed so they could run an electrical line under the slab. Before you pour, plan ahead for any future needs.

RAIN GUTTER DRAINS

Most of us living in regions with wet winters have rain gutters along the eaves of our homes. Although the drips are funnelled into one spot, the runoff can cause other annoying problems such as flooded planters (Fig. 13-1).

By simply running a section of 3-inch plastic drainpipe under the concrete, you can direct water anywhere you want (Fig. 13-2). The pipe is relatively inexpensive and is available at lumberyards and hardware stores. Concrete can be laid against it with no added difficulty. Flexible pipe eliminates the need for connecting joints (Fig. 13-3). You'll have to dig a 3-inch-deep trench under the slab area. This will ensure a full 3½ to 4 inches of concrete

over it. The other thing to remember is the slope. Gutters attached to the house generally should drain into a pipe going away from it—just like the slab. If you follow the slope of the formed area, you won't miss. Use common sense in the layout and plan for the water exit. If necessary, run the pipe out to the front curb. This will prevent any water buildup in the yard.

Solid or flexible plastic pipe can be used. The pipe usually will not have to directly connect to the drain spout (Fig. 13-4). Positioning the pipe opening directly under the spout will let the water drain into it. Use any type of pipe you want, but it should be at least 3 inches in diameter. The need for such a size is to prevent clogs. Pine and Douglas fir needles, leaves, grass clippings, and other material can plug the pipe. By providing a large diameter, these possible obstructions will be able to flow through the pipe and not get stuck.

WATER SUPPLIES

One of the more common oversights is failing to

Fig. 13-1. Rain runoff from gutter spouts can quickly fill a planter. Plastic drainpipe is inexpensive and easy to install under your walkway.

Sleeves provide access for more than just sprinkler system pipe. They can be used as guides for anything you might want to run under the concrete: service lines for heat pumps, drains for outdoor sinks, etc. Try to visualize your yard's landscape when it is completely done. See if you'll need any lines that would end up crossing the concrete. By doing this, you'll have the ability to complete a project with no worries of having to tunnel under concrete.

If a sprinkler system is already in use, you can protect the pipe from damage by installing supply lines under the concrete (Fig. 13-7). Power lawn edgers are notorious for cutting and breaking plastic sprinkler pipe and sprinkler heads. By laying the supply pipe just inside the concrete, it will be protected from the edger blade. In very cold climates, the concrete can even act as a blanket to prevent water from freezing. You will have to drain the system before winter, but any water left in the pipe might stay in a liquid form if it's under the concrete. This might prevent the water from freezing and cracking the pipe.

ELECTRICAL LINES

How many times have you wished that electricity could be provided to a shed or workshop without the wires being strung from the eaves of the house and over the yard? By laying a section of conduit or plastic pipe under the concrete, you can fish wires through at any time. You can even provide electricity to a particular spot in the concrete. In Fig. 13-8, the homeowner has run a "dead" electrical line to the center of his front walkway. The metal can covers a pipe elbow and some extra wire. Later he will install a small water fountain in the middle of the large front walkway. Because he is not sure when the fountain will be installed, the can will be covered with concrete. By taking very accurate measurements, he will be able to pinpoint the spot where the can is, break out the 1½ inch of concrete over it, and pull out the wires. The walkway looks perfectly fine without the fountain. When the fountain is put in later, the electrical source will be there.

provide lawn sprinkler supply pipe. Getting water from the faucet to a grass area across a concrete walkway is difficult after the concrete has been poured. You have to use special water applicators to tunnel a way under the walk. Therefore, why not place a piece of sprinkler pipe under the walkway before you pour (Fig. 13-5)? If you have no immediate plans to install a sprinkler system, a sleeve can be installed as a guide for future pipe (Fig. 13-6).

Fig. 13-2. When planning to put drainpipe under the slab, be sure to allow time to dig a trench under the base so that a full 4 inches of concrete can be poured over it.

Another very important electrical line can be supplied to the side of a hot tub or spa. It is very dangerous to put electricity near water, but this wire will connect to an emergency shutoff to the spa pump (Fig. 13-9). Occasionally people have become trapped by the suction drains on some powerful suction pumps to spas. It is usually because a protective cover has been removed from the drain.

Nevertheless, being able to shut down the pump immediately during an emergency might be a lifesaver. Consider this option if you are going to pour a concrete deck around your spa.

If you are planning a patio slab, an electrical outlet might be appropriate at one end of it (Fig. 13-10). This outlet could provide electricity for televisions, radios, or other items you use during

Fig. 13-3. Flexible drainpipe does not need special connectors.

summer barbecues. You may even want to supply a natural gas line to your gas barbecues (Fig. 13-11).

When installing open-ended pipe under your slab, be sure to allow at least 3½ inches over it for the concrete. Cover the open end of the pipe with heavy-duty duct tape.

Any concrete that falls into the pipe will plug it up. If you have to attach additional pipe to the one in the concrete, leave enough of the pipe sticking up out of the concrete so that later connections can be easily made. When constructing a job using electrical and natural gas lines, use caution. Check with the building department in your area for guidance. Electrical and gas connections must be done according to certain building codes. Violation of these codes could result in a fine, or they could create a fire hazard. Check with the local authorities and be sure your job is safe.

FOOTINGS

The only time you will need to install footings is when a structure will be built on the concrete. Footings are holes in the ground that will contain concrete and provide support for the structure on top it. Different areas have various specifications for each type of structure and the footings under it. Your local building department will have those figures. Most structures will also require a building permit. Again, the building department will assist you.

Solid footings running along the entire perimeter of a slab are normally required for room additions and garages. These footings will be about 12 to 18 inches wide and deep and will be placed under all exterior walls (Fig. 13-12). Rebar rods also must be installed in these footings. The concrete placed in the footing will provide a solid sup-

174

Fig. 13-4. Solid drainpipe can be used as a basin for gutters.

port for the wall. The rebar will add strength to the concrete and prevent cracks. The dimensions for footings vary from building department to building department. Contact them for assistance regarding room additions and garages, the building department will want to inspect your footings before the concrete is poured. Make allowances for that when scheduling the concrete delivery.

Patio covers generally require a much smaller footing—sometimes called a pier footing. A hole measuring either 12 by 12 inches or 18 by 18 inches is needed under each patio cover support post (Fig. 13-13).

STIRRUPS

Stirrups or wet post anchors are metal brackets that fit halfway into the concrete. The portion above the surface will be used to secure the bottom end of the supporting post to the patio cover. They come in sizes to accommodate 4 × 4 , 4 × 6, and 6 × 6 wood posts (Fig. 13-13). Placing the stirrups is critical. If they are not in a straight line, construction of the patio cover will be hindered. Because the posts will support a large header, they must line up. If not, the post will not be able to sit vertically upright and correctly support the header.

To ensure accurate placement of the stirrup, use nails and string for guides. In Fig. 13-13, a nail sticking out of the form at the top of the photo designates the center point for the stirrup. The string at the bottom is the guide used to keep all stirrups in a straight line. These points must be determined before the pour. Using the guidelines provided by the building department, you can locate the points for stirrup installation. On the end form, place a nail at each point that a stirrup is needed. On

175

Fig. 13-5. Planning for a sprinkler system ahead of time is a good idea. Tunneling under concrete is hard work. Place a sleeve under the walk when you pour it.

176

Fig. 13-6. Large sleeves can be placed under concrete for future water or electrical lines.

the side forms, place a nail on the side for string attachment. The string will determine how far away from the end of the concrete the stirrups will be placed. This distance should be far enough away from the edge of the concrete, so that an edger tool can be used between the stirrup and the form. Depending on the design, I normally place the stirrups about 1 foot from the end form.

To determine the actual distance between stirrups, refer to the building department's guidelines. They generally should be no more than 10 feet apart and can allow for an overhang of 3 feet on each side. If you were going to erect a patio cover measuring 36 feet across the back, you would need three stirrups. Going from a side form, there would be one at 3 feet, the next at 13 feet, then 23 feet, and finally at

Fig. 13-7. Sprinkler system pipes are better protected when laid in concrete.

33 feet. The final 3-foot overhang will bring the total to 36 feet. These figures are determined by the dimensions of the cover and the type of material used. The size of the lumber and the spans it will make determine stirrup placement.

Stirrups can only be used for patio covers and open carports. The only stability given by them is the sideway movement of the bottom post (Fig. 13-14). If you plan a room addition or garage, you'll need support for the entire wall. After the concrete has cured, a 2 × 4 will be laid on the edge of the concrete, and the wall will be attached to it. The 2 × 4 plate will be anchored to the concrete with the use of J bolts (Fig. 13-15). J bolts are inserted into the

Fig. 13-8. Providing electricity to an area of the slab is a good idea. Making notes as to the exact spot of the electrical line allows you to cover it with a couple of inches of concrete until you decide to break it out and use it.

concrete after the tamping and first bull float. They are generally placed about 4 feet apart in the middle with one at each corner. The bolts are sunk 1 foot in from the actual corner.

The building department will give you guidelines to follow the adequate placement.

When placing J bolts, you'll need to use a small piece of 2 × 4 as a guide. Nails on the sides of the forms will locate the places to put them. With the piece of 2 × 4 laid on the concrete, push the bolt down into the concrete. Leave enough sticking out over the top of the 2 × 4 so that a washer and nut can be attached. With the J bolts extending out from the top of the concrete far enough, you will be able to bolt down the wall in place.

REINFORCING WIRE

Reinforcing wire or hog wire is 10-gauge steel made into wire with 6-inch by 6-inch squares (Fig. 13-16). The rolls are 7 feet wide and 100 feet long. You can buy as many feet as needed, but you shouldn't have to buy an entire roll. The wire is available at concrete plants and lumberyards.

Hog wire adds some strength to the concrete. This wire in a slab usually will prevent cracks from becoming hazards. If a crack does appear, the hog wire will keep the level of the concrete steady. It will not allow one side of the crack to rise higher than the other. Essentially, hog wire will hold the slab together (Fig. 13-17). Hog wire usually is not needed. Slabs that will be used only for people will not have the need for strength reinforcement. It is mainly a heavy-duty additive for most patios (Fig. 13-18). I recommend that all driveways have hog wire. The weight of automobiles, trucks, and possibly delivery vehicles will add to the stress that the concrete is under. Hog wire will add strength.

Placing the wire is simple. Because the 7-feet width is standard, lay it down in a way that will reduce overlapping. If your driveway measures 12 by 28 feet, use four 12-foot pieces. Four 12-feet lengths will completely cover the area, and there will be no wasted overlaps.

Lay the wire so that the natural curl will be

Fig. 13-9. A spa emergency cutoff switch next to the sides of the spa could be a lifesaving feature.

180

toward the ground. Because the wire comes in large rolls, it will tend to roll back on itself after it is laid down. Flip the wire over when placing. The ends will push themselves into the ground. You can hit the wire in different places with a shovel to force it down. Wire sticking up will force its way through the concrete. If allowed to do that, you will find pieces coming through the cream while you are finishing.

You'll want the hog wire to lay flat, but not completely on the ground. During the pouring phase, you will have to periodically reach down and pull it up to the middle of the concrete (Fig. 13-19).

Some concrete finishers like to place rocks under the wire at different points, ensuring that the concrete will fully surround it. I have found on large pours that the wire is a nuisance. Someone always seems to trip over it. I leave it flat and just pull it up during the pour.

Hog wire sells for about 50 cents a lineal foot —actually 7 square feet because it is 7 feet wide. This low cost will ensure a crack-free driveway if that base is solid and evenly graded. Hog wire provides excellent protection for driveways and slabs that will be subjected to a lot of weight.

CONCRETE SPLATTER PROTECTION

Concrete work is very messy. No matter how

Fig. 13-10. Providing electricity at the end of the patio will be perfect for watching television or listening to the radio during summer barbecues.

Fig. 13-11. Some gas barbecues can operate on natural gas. A gas line run to a corner of the patio might come in handy.

Fig. 13-12. A footing this size is normally required for a room addition or an enclosed garage.

messy. The cream will splatter everywhere. Newspapers or plastic will protect the side of your house from the splatter. While you are finishing, you can tear off the covering. Helpers will be needed to pull it away from the wall while not letting it fall on the concrete. Any concrete residue that falls on the slab can be finished into it (Fig. 13-21).

Be very concerned about the plastic's position. Do not allow the bottom end of it to be in the concrete. You might have to chalk a line where the top of the concrete will go. If the end of the plastic is allowed to fall into the concrete, you'l have a tough time getting it out. You will have too much repair work to do afterward. When taping the plastic in place, use heavy-duty tape. Duct tape seems to work quite well. There is nothing worse than having the plastic fall down in the middle of pouring.

Besides the wall to the house, there might be other things to protect. Railings, metal patio cover posts, and pool sumps should be protected (Fig. 13-22). Use heavy-duty tape or tape plastic strips around projections. Concrete can be taken off many surfaces after it has hardened, but this is extra work. Using a little forethought will protect the things you want to and make cleanup easier.

WETTING THE BASE

Before any concrete is placed on the ground, the base should be damp (Fig. 13-23). The ground will probably already be damp during winter. Cool and damp weather will not cause the slab to "flash." The base should be damp so that the water in the concrete will not evaporate too fast and make finishing difficult. During wet weather, make sure the base is only damp. If it is by itself do not add any water. A base that is too wet will keep the water in the concrete too long, and the slab will take hours to set up.

Be sure the base is soaked during summer months. Apply enough water to make small puddles. Get that base as wet as possible, especially during heat waves. If the base is only damp, the water from the concrete will immediately drain into the ground. This condition will force the concrete to flash. It will get hard so fast that you might not be

careful you are, concrete will always get splattered on something. You should always put something up on the house (Fig. 13-20). You can tape newspapers or plastic against the wall. Protection of the wall for at least 2 to 3 feet from the top of the slab is about mandatory. Concrete falling from the wheelbarrow or the chute will splatter. Tamping is extremely

Fig. 13-13. For patio covers, the only footing required in most places is a simple pier footing. The stirrup being supported by a 2 × 4 (just to take the picture) will be sunk halfway in the concrete and will serve as a support for a patio post.

Fig. 13-14. The wavy arms at the bottom will go into the concrete. The straight arms on top will support the 4 × 4 post.

able to force the tamp down hard enough to make a mark.

This flash condition will ruin a concrete job. Because of water loss, an adequate tamp will not be possible. The float and fresno will be useless. You will have to sprinkle water onto the slab and immediately begin finishing. I have gone through this experience. Ending up with bloody and blistered hands, I was able to finish the slab to a decent condition. If I had more helpers and if the base would have been wetter, the problem would have never occurred.

Have plenty of help and really soak the base during hot days. If the ground is extremely dry, let a sprinkler soak the entire area for a few hours. In hot weather you can pour the concrete on top of mud, and it will still set up in a few hours. To be sure of base wetness, ask the concrete driver for his opinion. If you're pouring on a hot summer day, have that base completely soaked.

CONCRETE PUMPING PREPARATIONS

If your concrete job is going to be pumped, be sure to have one helper available to pull hose. Instruct him to avoid kinking the hose. If stringers are used in the design, make certain that the ones that will have the hose across them are securely supported and braced (Fig. 13-24).

Have a place for the water, sludge, and possible friction-reducing soap agent to go. The first

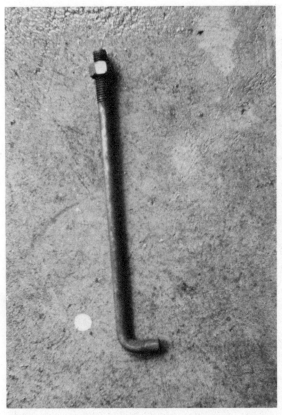

Fig. 13-15. J bolts are used to secure permanent enclosed walls.

ground at the furthest point of the slab. You'll know when the mud is about ready to come when you hear a definite "swishing" sound. That sound is made by the pea gravel scraping against the side of the hose. When it starts to come out the hose, move to the place you want concreted first. By holding the end close to the ground, the concrete will not splatter.

Fig. 13-16. Hog wire adds strength to a slab and prevents cracks from enlarging and forming lips on the surface.

thing to come out the end of a concrete pump hose is water. The pump man will fill the hopper full of water to prime the pump. That water has to go somewhere, and you do not want it inside the forms. The end of the hose should be stretched out far enough to reach the furthest point on the slab. Extra hose may be needed to reach past the forms so that the water can be drained. If necessary you can bring back the hose to a closer location for water drainage (Fig. 13-25). Make sure the water will not enter the formed area. The extra water will dilute the concrete and cause that area to stay wet much longer than the rest.

During the first few minutes of the pump job, you may have to man the end of the hose because the pump operator will be making sure that the pump is running smoothly. After the concrete starts coming, place the end of the hose close to the

Fig. 13-17. Hog wire will hold a slab together.

Fig. 13-18. Hog wire inserted in a patio slab is a heavy-duty luxury.

Fig. 13-19. Wire placed flat on the ground will have to be occasionally lifted into the concrete during the pour.

Fig. 13-20. The most practical thing a concrete man can do is place plastic along the house to protect it against splatter.

Allow only enough concrete to fill the area to the top of the form. Any more will only cause more screeding work. The pump operator will take over in just a few minutes. Be sure to talk over the job with the pump operator. Let him know exactly what you have in mind, that you are a novice concrete man, and that you want him to man the end of the hose and place the concrete.

PREPARING FOR THE SECOND LOAD

Although I recommend that a novice concrete finisher attempt jobs no greater than 3 to 4 yards, those of you with at least some experience might be doing a job that requires two loads. Getting ready for the second load includes several things.

You should have enough helpers so that at least one of them can begin tamping the first load right away. While the second batch is being poured, the first should be getting tamped and bull floated. Waiting too long to tamp and float will result in the first load flashing.

At the end of the first load, rake the end of the

Fig. 13-21. This photo shows how much concrete can get on the wall if it is not protected.

concrete to form an even slope. If you have to wait for the second truck, make a lip at the end of the concrete about 1 inch deep. Rake the rest out in an even slope. When the fresh concrete is placed, there will be no 4-inch drop-off of the first load for it to mate. If allowed, this edge will form a cold joint, and I guarantee that a crack will form there.

By sloping the concrete away from a 1-inch lip, the new mud will have somewhat of an even base.

Closer to the 1-inch lip, both the first and fresh concrete will be able to mix after being tamped (Fig. 13-26).

When pouring a job that requires more than one load, you must have plenty of help. You'll actually be pouring two slabs. While three or four workers are unloading the second truck, the first load is setting up. You must have at least one worker available to start tamping right away on the first

Fig. 13-22. Use heavy-duty duct tape to protect pool railings, patio posts, and railings.

Fig. 13-23. The base for concrete must always be wet down before the pour, especially during summer.

load. If you are still pouring after the tamping is done, that worker should start to bull float. If you don't have enough help during a large pour, you will work terribly hard, and you stand a good chance of "losing" the slab. Losing the slab refers to concrete that flashed and didn't get a good finish.

STRINGERS

In preparation for a pour using stringers, make sure that long stringers are properly staked, supported, and braced (Fig. 13-27). You may have to place stakes on each side of the stringer across from each other. By sandwiching the stringer between two stakes, you can be assured of its support.

Be certain that stringer joints are straight and square (Fig. 13-28). Placing a toenail into the notched joint will make doubly sure that it is secure. Don't place a nail through the top of the joint. That

Fig. 13-24. When pumping a job around stringers, be certain each stringer that has to support the hose is well-supported itself.

Fig. 13-25. You can double back the pump hose to a location for draining pump priming water.

Fig. 13-26. When preparing for the second load of concrete, rake the edge of the poured mud at a slope with a 1-inch lip at the top.

Fig. 13-27. Before pouring around long stringers, be sure they are properly supported.

Fig. 13-28. Be sure stringer joints are square and straight.

Fig. 13-29. Don't forget to tell everyone that you don't want concrete in planters.

nail would show and make the job look bad. If the grade is a bit too low, place dirt under it or drive a short stake down the side of one form and nail it. If you are staking and nailing, be sure the top of the stake is at least 1 to 1½ inches below the top of the stringer.

If planters are in your slab, don't forget to tell all the helpers about it (Fig. 13-29. Don't fill a planter full of concrete. It would be a waste of concrete, and that concrete will be hard to remove.

Chapter 14

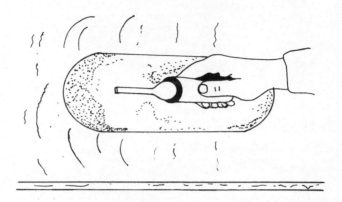

Pouring

THE MOST PHYSICALLY DEMANDING PART OF concrete work is getting the concrete from the truck to the forms and screeding it off (Fig. 14-1). Your goals are to make this chore as easy as possible and, of course, to do it right.

CONCRETE COMPANY

I have had very good luck with both large and small concrete companies. Smaller companies have delivered most of my concrete.

Large companies deliver concrete to many contractors who pour lots of jobs. They are always busy and are on a tight schedule. The smaller ones, even though they do serve contractors, have more patience with the novice finishers. Their overtime rates generally allow more time per yard to unload, and the drivers are more familiar with the needs of the homeowner (Fig. 14-2).

Smaller companies appreciate your business just like the bigger ones, but they must work a little harder to gain more accounts. For this reason, they will bend over backwards to help you. They might spend more time with you in assisting with calculations, mix, and ideas on how to make the pour easier. By spending more time with you and being of a little more help, a smaller company may be best.

I am not recommending that you stay away from large concrete companies. The smaller ones, though, might have more available time to serve you and your needs better. The best thing for you to do is to contact the concrete companies in your area and talk to the dispatcher. The feeling you get afterward, along with the prices and overtime rates, will be all you need to make a decision.

TRUCK DRIVER

When the concrete truck arrives, talk to the driver. Explain the job site and what you have planned. Show him the formed area and ask if he sees any problems. Listen to his advice and make changes if necessary. You may have to place a few more stakes, brace a form, or build a better ramp for the wheelbarrow.

Fig. 14-1. When the chutes will not reach, you will have to move the concrete with a shovel or rake.

Dumping the concrete into wheelbarrows will mean that the truck will park in one location and stay there. The driver will operate the concrete flow from controls at the rear of the truck. The same method holds true for pumping.

If the concrete will be placed directly from the truck to the forms, the driver will operate and maneuver the truck and, simultaneously, control the vertical movement of the chute and the flow of the concrete. Controls inside the cab allow the driver to operate everything.

Because the driver will be in the cab, he can only see the job from the side mirrors (Fig. 14-3). Therefore, someone must be within his sight to direct him (Fig. 14-4). Because the engine on the truck will be running at a high rpm, the driver will not be able to hear you. It is very important to establish hand signals before the pour.

Normally by holding a clenched fist in the air, you can signal the driver to stop the truck (Fig. 14-5). You should do this when the truck is in position. After the chute has made a complete arc

Fig. 14-2. Smaller concrete companies generally allow more time per yard to unload.

and the right amount of concrete is on the ground, the driver will pull the truck forward. Stopping him about 2 feet ahead will ensure that you get the concrete filled in to the last part poured. When the arc is completed, wave your open hand under your neck to stop the flow of concrete (Fig. 14-6). This will tell the driver that you want the flow to stop; enough concrete is on the ground. This signal is very important. Too much concrete on the ground means that you will have to move it with shovels. That will be a delay in time and much extra work.

Because a wet mix of concrete flows easily, you should shut down the flow early. This will let the mud in the chute to come out where you want it, and you won't have to stand there with a shovel holding it back in the chute.

The end of the chute is vertically controlled by a hydraulic ram. Controls in the cab allow the driver to raise and lower it. During the pour, you should have the end of the chute about 1 to 1½ feet off the ground. To have the driver adjust it, simply point up or down. Point to the chute first and designate that it is what you want moved (Fig. 14-7). When it is in position, point to the ground. That will be the signal to let the concrete flow.

All of these signals are simple. There is no real standard set for driver signal operations. By talking to the driver first, both of you can get a system set. If there is ever a problem, stop the truck and again talk to the driver. Work together and the job will go smoothly.

FINAL CHECK

While you and the truck driver are walking the job, make some final checks. Kick the forms to be sure they are braced and supported. If the job is going to be pumped, be certain that all forms where the hose will cross are secure (Fig. 14-8). Be sure that any pipes extending out of the slab area are braced and the openings covered (Fig. 14-9).

If hog wire is laid, make sure it is flat. No piece of the wire should be allowed to stick up past the tops of the forms. Also, notify all the helpers that they'll have to occasionally reach down and pull the wire up into the concrete. Wire flat on the ground

Fig. 14-3. The truck driver cannot see directly behind the truck. It is important to have someone guiding him when backing up.

will not add strength. It must sit in the center of the concrete.

Finally, be absolutely positive that the base is wet. I can't stress too much how vital this is during hot weather. A dry base combined with hot, dry weather will guarantee a lost slab. Water in the concrete must remain for some time. If the dry conditions exist, the slab will flash and make finishing impossible.

On a direct pour, the driver will have to see the area in which he'll have to drive. Both of you should be aware to how he is going to maneuver the truck. Some jobs might require the truck to pour half of the job and then completely reposition itself. While working under this situation, advise the helpers when the truck will have to be moved. At that time, one man should be in the front of the truck to guide the driver. Another man will have to be at the rear; he will steady the chute and guide the driver. Be very careful to stay clear of the truck. With all of the engine noise, the driver will not be able to hear someone yelling. Be very cautious to have nobody

Fig. 14-4. Someone should always be in the driver's sight on either side of the truck.

Fig. 14-5. A clenched fist held up signals the driver to stop the truck.

at the back of the truck who is out of the driver's sight. When he backs up, he doesn't want to run over anyone.

DIRECT POUR

After the truck is in position, the driver will get out of it. He probably will ask you if you want to see a sample of the mix. If he doesn't ask you, you had better ask him to see a sample.

The driver will let a little concrete come out of

the drum. Then he will stop it. Look at the mud and see if it is the right consistency (Fig. 14-10). If it is too stiff (dry), have the driver add water. Stiff concrete will not flow. It will batch up and not come down the chute easily (Fig. 14-10). At that time, you should have the driver add 1 gallon of water for every yard of concrete in the drum. He will start spinning the drum, adding water at the same time. The motion of the drum will mix the concrete and water. When he is finished, look at another sample.

Fig. 14-6. Waving an open hand under the neck signals the driver to stop the flow of concrete.

A perfect mix is one that flows easily. For the most part, you should work with a wet mix. It is easier to place and much easier to screed (Fig. 14-11). If you are not sure of the mix, ask the driver's opinion. When you think the mix is correct, dump some on the ground and see how it goes (Fig. 14-12). The mud should not pile up on you; it should flatten itself out and flow outward. A good mix of concrete will fall off the chute in a constant pour. A dry mix will plop on the ground in batches. A solid stream of concrete coming off the chute is a guarantee of a nice wet mix.

If during the course of the pour you think the concrete is starting to get dry, stop the driver and have him add water. There is no set limit as to how much water can be added. To be on the safe side, though, only add 1 gallon per yard at a time. This will protect the mix from getting too wet.

Dumping the concrete directly from the chute requires the use of at least three and preferably four helpers. One will operate the chute. He will be responsible for moving the end of it in such a way as to get the concrete on the ground in the places it is supposed to be. He should be keenly aware of the height of the mud. Too much will make screeding difficult, and the mud will have to be moved by shovel or rake. By slowly keeping the chute in motion, he can move it from side to side by making long and steady sweeps (Fig. 14-13).

The chute man will also be responsible for signals given to the driver. Because the driver has a mirror on each side of the truck, he can see you through either of them. He will watch the chute man for directions. When things are going right, the driver can tell when he will have to move forward. Generally, at the end of each arc, he will move the truck ahead slightly. When the truck moves, the chute man can start the chute back the other way. The chute man should always try to pour an even amount of concrete. Using forms and screed forms

Fig. 14-7. Pointing up or down refers to the up or down movement of the chutes.

as guides, he must try to get just 4 inches of mud on the ground. This takes practice (Fig. 14-14).

During the pour, *never* be afraid to stop the truck. If too much mud is on the ground, stop the truck and move the concrete as needed. If you don't, you'll be working very hard to move the excess concrete later. It is far better to stop the truck, catch up right away, and then move forward. Getting the mud on the ground and screeded is hard work. Take your time and do it right. This part of the job is not a time to worry about overtime. If it takes a little longer than you expected, so be it.

Moving the concrete once it is on the ground is necessary. I don't think anybody can lay down a perfect grade right from the truck. The chute man can sometimes help by pushing the concrete with his foot (Fig. 14-15). Sometimes the end of the chute will not reach into a corner. By using his feet, he can move small amounts into corners and help to spread the concrete.

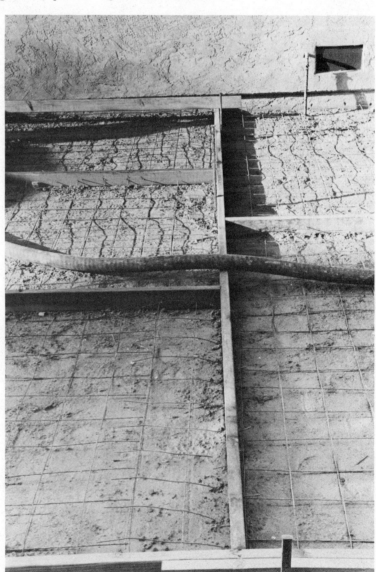

Fig. 14-8. Forms that will be subjected to added weight or disturbance factors must be beefed up with stakes.

Fig. 14-9. When pipes extend out of the slab, be sure they are secured for the pour.

Using a rake or shovel (a rake is better), the screed men can accurately place the concrete. High spots will have to be pulled back, and low spots will have to be filled in (Fig. 14-16). Before enough concrete is on the ground to screed, these workers can help level out the concrete. Getting the level of the concrete close to grade with the rakes will make screeding much easier. Use the rake just like you would to smooth a flower bed. Get the concrete as flat and even as possible (Fig. 14-17). After about 4 feet of mud is on the ground, stop the truck and screed. At this point, another helper or the chute man can use a rake or shovel to pull concrete away from the screed board or fill in low spots (Fig. 14-18).

Working together as a team is most crucial. The screeders, chute man, and extra helper must talk to each other. Screeding will guarantee a flat

Fig. 14-10. Always ask to check the mix. This mix is dry. Note how it piles up in the chute rather than flowing freely.

Fig. 14-11. This is a good mix. The concrete flows evenly—almost like a liquid. Note how it flattens out when it hits the ground.

Fig. 14-12. A good mix of concrete will not pile up on itself like a pyramid. It should spread out by itself—much like a thick salad dressing.

slab. Unless all people work together, some low and high points might present themselves. The screed men should let the shovel man know where they want more concrete or where they want some removed (Fig. 14-19).

Use guides along the job to help measure accurate concrete depth while pouring (Fig. 14-20). Stop the truck as necessary, especially after the chute is about 4 feet ahead of the screed men (Fig. 14-21). Take the needed time to do the job correctly.

WHEELING

Placing the concrete with wheelbarrows takes a lot of energy. Concrete is very heavy, and it takes some time getting used to wheeling it. Position the truck as close to the job as possible, making the wheel as short as can be expected. On delicate jobs with redwood benderboard, you might want one

Fig. 14-13. The chute man should constantly be moving the chute. An area can be covered slowly to a depth very close to 4 inches.

Fig. 14-14. The chute man can use screed forms and the concrete on the ground as a guide for the 4-inch depth.

Fig. 14-15. Many times wet concrete can be pushed into place with your foot.

Fig. 14-16. A heavy-duty rake is excellent for moving mud.

Fig. 14-17. Rough screeding the concrete with a rake makes actual screeding easier, because there will be less excess to move.

Fig. 14-18. While two men screed, the shovel man removes excess concrete and fills in holes.

man to wheel the concrete to the area and then fill in certain spaces with a shovel. This is most commonly done when pouring steps (Fig. 14-22).

Tell the truck driver to fill your wheelbarrow only half full. This will let you get a feel for the weight and maneuverability of the wheelbarrow. It will also allow you to determine the best route and get used to ramps. After you have gotten the feel for

wheeling, you can have the driver fill it with more concrete. Many helpers have started out with full wheelbarrows only to lose the load halfway to the forms. None of us want those types of accidents. Take your time and get used to the work. Get all the concrete inside the forms and not all over the yard.

When you dump the concrete out of the wheelbarrow, use a quick throwing motion. Too many

Fig. 14-19. Communication between all the workers is vital during the pour.

Fig. 14-20. The screed form in the middle of the slab makes a good guide for concrete depth.

wheelers slowly tilt the wheelbarrow and allow the concrete to gently fall out of it. This action makes a mess. The mud will fall and splatter everywhere. By rapidly tilting the wheelbarrow, the forward edge of it will almost touch the ground before any concrete comes out (Fig. 14-23).

Start to tip the wheelbarrow about 2 to 3 feet before the edge of the concrete already on the ground. The quick action of the wheelbarrow being tilted will force the concrete ahead, filling in the empty space. All of this should be in one motion. When you come up to the area needing concrete,

use your forward momentum to help tilt the wheelbarrow. By throwing your arms up, the handles will quickly move up and the front will go down. The mud will be forced out of the wheelbarrow and run into the concrete in front of it. This takes some practice. Just try to remember to dump the load quickly and all in one motion. This will greatly reduce splatter and help spread the concrete.

BLOCKS

Two-by-four wood blocks used to keep redwood benderboard in place should not be removed until

215

the concrete is in place. Placing the concrete inside areas formed with benderboard sometimes requires using a shovel (Fig. 14-24). Because the forms are somewhat fragile, the gentle placing action of the shovel will not disturb them. After some concrete is placed in one section, some should be added to the next. This will allow the weight of the concrete to be placed against the forms slowly and evenly (Fig. 14-25). Too much weight at one time will bow the forms.

Once the area is filled, the blocks can be gently removed (Fig. 14-26). You will have to fill in the space created by the block. Using the shovel, fill in the hole and mix the concrete with the rest. This action will make sure that the mud is blended together, reducing the chances of a cold joint and crack.

Blocks and support forms must be removed the same way. After concrete is poured on all sides of them, they can be pulled. More concrete must be

Fig. 14-21. Place about 4 feet of concrete on the ground and stop the truck. Screed the job to that point. This will keep you ahead of the job.

Fig. 14-22. To place concrete inside delicate forms, fill a wheelbarrow with concrete and shovel it from the wheelbarrow to the forms.

Fig. 14-23. When dumping a wheelbarrow full of concrete, tilt it up and dump the load quickly. This will reduce splatter and force the concrete out in a flat pattern.

Fig. 14-24. Use a shovel to place concrete inside delicate forms.

placed to fill the voids and mixed with the surrounding material. On long 2 × 4 support forms for expansion joint felt, you might have someone push concrete under the board while you are pulling it up. This will ensure that the felt stays straight and is not pushed over by the weight of the concrete on the other side of it.

Again, when using hog wire, you must remember to pull it up during the pour (Fig. 14-27). Blocks have to be pulled after the concrete is in place. Wire must be pulled during the actual placing of the mud. Many finishers forget to do this task.

The wire only has to be pulled up about 2 inches. The chute man pulls it up as he walks along, or the shovel man can stay in front of the chute and make sure the wire is off the ground.

PUMPING

Concrete pumps get concrete from the truck to the forms the easiest way. Concrete is dumped directly into the hopper, is forced through the hose, and comes out where you want it.

Pumps are generally owned and operated by independent businesses. The pump man will ad-

Fig. 14-25. Put the concrete into benderboard forms slowly and evenly.

vertise his services in the newspapers and the telephone directory's Yellow Pages. They will be listed under concrete pumps. Some concrete companies have their own pumps. Either way, the concrete dispatcher can put you in touch with a pump man if you don't want to call directly.

On the job site, the pump man will first look at the area's general outline. He will determine where to spot the pump and how he wants the hoses to go. He makes allowances for the concrete truck and positions the pump for access by the concrete truck (Fig. 14-28). After the pump is positioned, he will start laying out the hose.

Concrete hose usually comes in 20 to 25-foot

Fig. 14-26. Remove blocks after the section is filled with concrete. Have extra mud ready to fill in the holes left by the blocks.

Fig. 14-27. Don't forget to pull up hog wire during the pour.

When the concrete truck arrives, the driver will be guided to the pump. He will back the truck to the hopper on the pump. After he looks at the job, he and the pump operator will check the concrete mix. When the consistency is right, concrete will begin to flow into the hopper. With the pump running, the operator will remain at the pump for a few minutes so he can check the pump and make sure it is running correctly. At that time, you will have to man the end of the hose.

The first thing to come out of the pump will be the water used for priming. Be absolutely sure that the water goes to the right spot. You do not want it inside the forms. After the pump is assured of run-

Fig. 14-28. Concrete pump spotted with concrete truck in position.

sections. The first section will be placed at the farthest end of the job. Subsequent sections will be attached by special couplings (Fig. 14-29). The pump operator will take care of the hose and couplings. You can help if you want, but it is not mandatory.

After the hose is connected, the operator will place a sponge inside the opening at the bottom of the hopper or at the point where the hose attaches to the pump. Sometimes, although not recommended, he will kink the hose at the pump (Fig. 14-30). This will be done so that water can be put in the hopper and act as a priming agent for the pump.

When the water is placed in the hopper, the operator will then string out the remote control wire (Fig. 14-31). This remote control will allow the pump to be shut down without anyone having to actually be at the pump. If the concrete starts to get ahead of the screeders, the pour can stop. When the pump stops, the concrete truck driver will stop the flow of concrete. The remote control switch is nothing more than a toggle switch. Either it will be placed at a strategic point by the forms, or the man in charge of pulling the hose can man it (Fig. 14-32).

Fig. 14-29. Concrete pumping hose coupling.

them could be a major problem and will cause an unexpected delay.

The most important thing to remember is to avoid kinks in the hose. The man pulling the hose must constantly be on the lookout for potential kinks. If the hose were to kink, the pump's relief valve would open. Resetting the relief valve will mean that the concrete will have to be shoveled out of the hopper. After that, it will take about 20 minutes to reset the valve. Again, avoid unexpected delays by being prepared.

At the job site, the pump operator will man the end of the hose and place the concrete. For the most part, he can put the concrete down in a very level fashion. You can have a worker help spread the mud with a rake (Fig. 14-34). As the job progresses, two men will be on the screeds, one will have to pull hose, and the pump operator will place the concrete (Fig. 14-35). Screeding should be very easy, because the concrete will be wet and will have been accurately placed. If the screed men get behind, however, stop the pour and allow them to catch up (Fig. 14-36).

Pumping concrete is a great way to go. Besides the ease of which the concrete is placed, you get an extra man to help. After the pumping is done, though, the pump man will have to clean up the hose and the pump. He won't be able to help you. Cleaning the pump and the hose will leave a small mess.

Fig. 14-30. Pump hose kinked to keep primer water in the hopper. This is not a good idea because it wears out the hose. During the pour, a kink like this will force open a relief valve on the pump, delaying the job at least 20 minutes.

ning right, the pump operator will take over the end of the hose and place the concrete for you.

The truck driver will remain at the truck during the pour. He will control the flow of the concrete (Fig. 14-33). When the pump is shut down by the remote control, he will stop the concrete. He will also make sure that no large rocks go into the hopper. Large rocks will plug the pump. Removing

223

Fig. 14-31. Pump operator preparing to stretch remote control wire.

Fig. 14-32. Helper pulling back excess hose and manning the remote control switch.

Fig. 14-33. Concrete truck driver will stay at the truck and operate concrete flow controls.

Excess concrete in the hopper could be pumped back into the truck, but small amounts will be left in the pump and hose. Washing the equipment will leave concrete, rock, and sand on the ground. Generally, the pump man will at least wash the mess to the curb. The ultimate responsibility, though, rests on your shoulders. If you can spare a man after the tamping and bull floating, have him go out to the street and clean it up. Concrete is much easier to wash away before it sets up.

SPECIAL CONCRETE PLACING

Placing concrete against benderboard or forms taller than 4 inches, requires a slow buildup rather than a full pour right away. Because benderboard can bow easily, you should place an inch of concrete along both sides first. Fill in an inch at a time until the concrete is all in position (Fig. 14-37). The same holds true for forms on steps. Because there are only a few stakes supporting the form, the added weight of the concrete can push them out (Fig. 14-38). By slowly adding concrete to the entire formed area, the concrete acts as a support. You must go slow with steps. You will probably have to place the concrete with a shovel. Don't throw the mud against the forms; that's taking a chance on them moving or bowing out (Fig. 14-39).

Fig. 14-34. A heavy-duty rake is also good for moving around a pump mix.

Fig. 14-35. When all workers communicate and work together, the job always goes smooth.

Fig. 14-36. Even on pump jobs, stop the concrete to allow screeders to catch up.

Pouring concrete in special areas requires you to use your head. Don't get in a hurry. Place the concrete correctly, taking into account the size and shape of the forms.

SCREEDING

Screeding or rodding off can be done in two ways—with the use of screed forms or without them. On jobs with narrow dimensions, one man can operate the screed on top of the forms (Fig. 14-40). On slabs wider than 3 feet, it is difficult for one man. He will have to pull back the screed and any extra concrete, and he will have to keep an eye out for low spots (Fig. 14-41).

Two men screeding with the screed board resting on the forms must work together. By sliding the board back and forth in a seesaw motion, you can push down rock and pull concrete back at the same time. If you just pull the board straight back, the tips of rock might stick up and create small holes. The

sawing motion pushes the rock down and cuts the excess concrete off the top. The extra helper should be behind you pulling away any excess mud (Fig. 14-42). With all three of you working together, screeding should go quickly and easily. On each end of the screed board, be sure the board remains in contact with the form. Letting the screed board ride higher than the form will create high spots. While using the seesaw method, you can feel the board in contact with the form.

Because you will be standing in the concrete, footprints will be made. After you move your foot, you will have to fill in the holes. The easiest way to do that is with your foot (Fig. 14-43). Drag some mud from behind you and tap it in place. When the hole is filled and to grade, go ahead and screed over it. Always be on the alert for low spots. They look like areas of the concrete that haven't been screeded. The general appearance of the slab after a screeding has a rather smooth look. Low spots are rough. Use a shovel or your hands to put more concrete into those areas. You should go back to them and screed again.

Sometimes it is necessary to screed two or three times. I don't mean that you'll have to go over

Fig. 14-37. Along benderboard, fill the concrete along the entire length, 1 inch at a time, until the 4-inch depth is met.

the job two separate times, but you might have to go over a section twice. After a 3 or 4 foot section has been screeded, be sure it looks right. If low or high points are visible, fill in or remove concrete as necessary. Reposition the screed board and screed

Fig. 14-39. Most steps require that you fill in the concrete with a shovel.

Fig. 14-38. Because of the weight, concrete should be slowly poured into step forms an inch at a time.

it again. In Fig. 14-44, the strip of concrete in front of the screed men is filled with holes. After the short section is screeded, concrete will be placed into the holes. The entire section will be re-screeded. Work with each other to assure a flat slab. Small shovelfuls of concrete can be added to the surface and smoothed with the back of the shovel or a hand float. The area must be screeded again to ensure its flatness (Fig. 14-45).

When stakes stick up above the forms, you will

Fig. 14-40. One man can screed small areas.

Fig. 14-41. A screed longer than 3 feet requires two men.

Fig. 14-43. Filling in holes with foot when screeding.

Fig. 14-42. Rake man pulling back excess concrete from the screed.

have to adjust the screed board. The best way to do it is to have the opposite end of the screed board go past the stake obstruction. The board will actually be pulled on only one side, making it rest on the surface at an angle. After the end goes by, it must be repositioned ahead of the stake—in other words, back a couple of feet to where it started. Then the man at the stake side can lift his end over the stake, moving any excess concrete with his hand (Fig. 14-46).

By first going past the stake on the opposite end, the concrete can be screeded along the area that the stake wouldn't allow. When the ends are repositioned, there will be only a very small area that wasn't touched by the screed board. That area will be a small triangle just next to the stake. You can smooth that spot with your hand. Any small inperfections will be removed with the tamp and the bull float.

When large obstructions prevent the screed board access, you will have to either use a smaller screed or maneuver the screed board back and forth to reach the concrete. In very small spaces, you can use the hand float to screed. You will have to depend on your eye for guidance (Fig. 14-47). Each circumstance is different. You will have to use your ingenuity for your special job. Try to use a screed board rather than doing it by eye. Professionals can eyeball a screed quite well.

Special screed forms can be added to the job to help screed. In some cases, it will be a lot of extra work. The outcome will be a flat slab, and that's what everybody wants (Fig. 14-48).

SCREED FORMS

If you cannot use the screed board on top of the forms, you should erect the screed form and use it. The extra work installing them will make the screed job easier and the slab as flat as possible.

Fig. 14-44. Rough looking concrete, which is in front of the finisher at the bottom of the page, is a low spot and will have to be filled in.

The only things you'll have to watch for are keeping the ear on the screed form and preventing the form from being pushed down. Unlike regular forms on the ground, screed forms are suspended in the air by stakes. Between the stakes, you can put too much downward pressure on the screed and force the form into the concrete. This will create a low spot on the slab. You will have to concern yourself with pulling back on the board and not pushing down.

After the job is screeded, you'll have to pull the screed forms. This can be done while tamping. Have a helper stand by the forms and take the board from you. Then, if necessary, he can hand you a shovelful of concrete to fill in the holes. Work the concrete into the hole with your foot, then tamp.

Fig. 14-45. Fill in low spots slowly and rescreed.

Fig. 14-46. Moving the end of screed board over the stake. The screeders must work together to effect a good maneuver.

Fig. 14-47. Hand floating a screed in a tight spot. Notice how workers have stopped the pour so that they can take care of the problem.

WET SCREED

You can use the wet screed method without too much trouble, especially along walkways. This operation requires some special skills that you will have to remember.

To begin with, chalk a line along the house where the top of the concrete will go. With the chalk line as a guide, attempt to place the concrete up to it. As a precaution, place a chalk line 3½ inches above the first. With the lower line as a guide for concrete placement, the top line will be a guide for the top of the 2 × 4 screed board. When the concrete is in place, use a hand float to push the mud up to the line. Smooth the concrete up to the line so that it is perfectly in place (Fig. 14-49). The top of the floated concrete will be used as a guide just as much as the upper chalk line. Go along an entire edge for about 8 to 10 feet, making the concrete level and flat (Fig. 14-50).

When an edge has been hand floated to the lower line, you are ready to screed. The screeder on the chalk line side will have to direct his atten-

Fig. 14-48. Special screed forms will have to be used on jobs where normal screeding is difficult or impossible.

Fig. 14-49. Using the hand float to wet screed a slab. The concrete must meet the chalk line and be flat coming away from it.

mistakes should be corrected. If the board is allowed to dip, go back and fill in the spot, then screed again. This type of screeding is done best on narrow walkways. Screeding a large slab this way is hard to do. It takes a lot of practice and experience. In Fig. 14-51, the concrete men might have been better off using a screed form. It would have been much easier, and the slab would have been guaranteed flat.

POOL DECKS

Screeding around pool decks is not hard. Generally, decking is narrow compared to a patio slab. You

Fig. 14-50. Wet screed guide floated in.

tion to the upper line and the concrete surface. He will not be able to look for low spots. Therefore, the other screeder and the shovel man will have to help.

Using a firm hand, glide the screed board over the concrete and follow the upper chalk line. With both those guides, the slab should be screeded with no problems (Fig. 14-51). You cannot use the seesaw motion in wet screeding. The end of the screed board against the house must stay against it during the screed. That and the concrete at its edge are the only guides that man has. You'll have to go slow and keep a watchful eye on the guides. Any

Fig. 14-51. Wet screeding the concrete. Notice how the screeder pays close attention to the position of his end of the board. The rough surface in front of the screed board signifies low spots and must be filled in.

Getting the concrete in the forms and screeded is the major job, and it is physically demanding and time-consuming. You will probably have that overtime factor in the back of your mind, too. Don't fret. Every concrete man around had those jitters on his first job. If you have followed my advice so far, you won't have any problems. If there is something particular about your job that I haven't covered, by all means seek advice. As a professional concrete finisher, I have given advice to do-it-yourselfers. Depending on the situation, my fee was extremely low. If I didn't have to travel to the job site, I usually

Fig. 14-52. Hand floating the concrete around a pool's edge will prevent excess screed concrete from falling into the pool.

might want to hand float the concrete right next to the pool. This will prevent excess concrete from being forced into the pool (Fig. 14-52). There are many concrete finishers who have fallen into pools while pouring a deck. Use extreme caution around pools and be sure of your footing. Don't slip and end up in the water.

When wheeling concrete into decking forms, try to empty the wheelbarrows in such a way as to not direct the mud toward the water. Empty the concrete alongside the pool's edge—not at it. You won't want to clean up excess concrete in the pool.

237

didn't charge anything. Don't expect that from all concrete finishers. Concrete is their livelihood. If you need help, get it. Concrete finishers guarantee only one thing—that the concrete will get hard. You don't want the concrete for your job to get hard before your finishers are ready. Read the text carefully, talk to the concrete dispatcher, have your helpers ready, and you won't have any problems.

Chapter 15

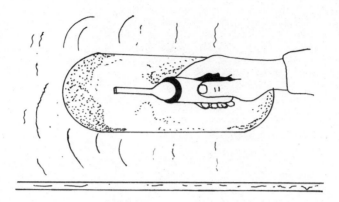

The First Hour

THE FIRST HOUR OR SO AFTER THE CONCRETE is on the ground and screeded will be a busy one. The slab will be prepared for finishing, stirrups or J bolts will have to be installed, and edges and seams will have to be effected. Some cleanup operations can start, but the emphasis must be on the concrete.

TAMPING

If you have enough help, one man can start tamping about the time the last bit of concrete comes out of the truck. With the exception of an exposed aggregate finish, I suggest that every slab be tamped—even one with a pea gravel. Some finishers find it unnecessary to tamp a pump mix (pea gravel). Simple bull floating usually pushes the rock down and leaves cream. Tamping this mix, though, will allow you more cream to finish. Finishing is hard enough without having little pebbles pop up after troweling. Sometimes that happens when you don't tamp pea gravel. Go ahead and tamp a pump mix. Assure yourself of a creamy surface to finish.

The tamper tool is a heavy-duty, close screened, wire mesh attached to metal brackets and a handle (Fig. 15-1). The wire screen forces down rock and brings up cream. Normally about 3 feet long, you will have to work back and forth across the slab to cover every square inch.

Unless the concrete is dry, and it won't be if the base is wet, you shouldn't have to use much force to push it down. The weight of the tamp alone is enough to produce the right effects. The surface should have an even appearance (Fig. 15-1). The cream will form a liquid type seal over the surface, leaving no holes or rock exposed. The bumpy wafflelike pattern will appear, and it is made by the holes in the wire mesh. When tamping has been done correctly, the entire slab must look like a watery surface with hundreds of bumps on it.

To aid in flattening the slab, tamp in the opposite direction you screeded. The screed may leave slight ridges as it is worked. By tamping along the line of those ridges, you will flatten them out (Fig. 15-2). The tamper need not be lifted off the surface more than a few inches (Fig. 15-3). You do not need

Fig. 15-1. The tamper in use on a slab.

to pound it into the concrete. A steady, even force is needed to push rock and bring up cream.

Because you will have to walk in the wet concrete to tamp, footprints will follow. This is why the tamp is designed to be used while you are walking backward. After you have stepped back, use your foot to tap the hole and bring concrete back to it. Tapping the concrete around the footprint will bring concrete toward the hole, just like when you played

with mud pies or sand castles. Patting the mud or wet sand brought up moisture and filled in holes.

After enough concrete has been brought back, tamp the spot. If the job has been done right, you'll never know there was a footprint in the slab (Fig. 15-4).

While using the tamp, keep the bottom flat. Don't allow one edge to hit the concrete harder than the other. This will cause divets in the surface and

Fig. 15-2. To help flatten the slab and remove screed ridges, tamp in the opposite direction of the screed.

will be difficult to repair. The tamper is balanced so that it will fall with the wire mesh flat. Let the tool do the work. Trying to force it makes the job harder, and you're more apt to end up with an uneven surface.

Don't forget to tamp the edges. The easiest way is to turn the tamp so that the 3-foot-long edge will go along the same lines as the form. Tamping the edges will ensure no rocks in the way when it

comes time to edge. Every part of the surface should be tamped—even the edges along stringers.

Timing is important. The sooner you tamp, the better. If you don't have a helper to start tamping while you are finishing up the pour, be sure to start as soon as possible. I would not let more than 10 minutes pass from the time you finish pouring to the time you should begin tamping. If the slab has started to set up, you will have to force down the

Fig. 15-3. On a wet slab, the tamp need not be lifted more than 4 inches to 6 inches off the surface.

Fig. 15-4. Footprints are covered and tamped because the tamp is designed to be operated while you walk backward.

Fig. 15-6. Having enough helpers means one on the tamp, another operating the bull float, and the third pulling screed pins and doing other assorted chores.

Fig. 15-5. The bull float smoothes the surface and brings up water.

Fig. 15-7. Waiting too long to float will allow the concrete to dry, causing holes and more work to smooth.

Fig. 15-8. Tamp and float the main body of the slab before placing concrete into time-consuming special forms.

Fig. 15-10. You can operate the float sideways if obstructions prevent you from pushing and pulling it normally.

Fig. 15-9. On wide slabs, you may have to float half of it from one side and the other half from an opposite side.

Fig. 15-11. Tamp and float the first load of concrete before the second load arrives.

244

Fig. 15-12. Enough helpers on the job can handle the first load, while others help to unload the second.

tamper. You'll have to pound the concrete hard enough to bring up water and fill in any gaps between the rock. If pounding doesn't do the job, you may have to sprinkle just a small amount of water on the concrete.

Using a very fine spray, very lightly dampen the surface. Putting water on the top of a slab is not good. It dilutes the cement solution and causes weak concrete. It is more vital that you bring up some cream. It is impossible to finish rocks. There must be a layer of cream on the top to finish.

By following instructions, this should not happen to you. Make certain that the base is wet. On hot days, soak the ground until puddles form. Pour over the mud. Have the concrete truck driver mix a good wet mix. During winter, this is normally not a problem. The ground is already wet from the rain and damp weather. During hot weather, though, be especially sure that moisture will stay in the concrete.

BULL FLOAT

To smooth the rough surface of your slab after the tamping, use a bull float. This tool combines a 3-foot-long piece of wood, an adapter, and the 6-foot-long extension poles. You should have enough poles to reach the furthest part of the slab. After the tamping, you should not walk on the slab.

The weight and surface area of the bull float will smooth the bumps and bring up more moisture (Fig. 15-5). Because the concrete will still be wet, lines and ridges will be left by the edges of the bull float. If the ridges are very prominent and the concrete is wet, you might bull float a second time about five minutes after the first. The surface will not be completely smooth. The fresno will do that. The bull float is the first step in getting the finish.

Bull floating can be started while the tamp man is still tamping. Sometimes, especially during hot weather, you must start right away. If enough help is available, one man can tamp, another can pull out screed forms, and the third can begin floating (Fig. 15-6). As far as bull floating is concerned, get it done as soon as possible. The tamping and bull floating will guarantee you enough cream to finish. I don't think any slab could be lost if the tamp and

Fig. 15-13. While pouring, use a hand float to clean off the tops of stringers.

246

holes with cream, you must immediately fresno. Although the surface will appear wet, the moisture will once again quickly evaporate. The fresno must be applied to completely smooth the surface.

On jobs that will require time-comsuming pouring applications, it is best to have the main body of the slab tamped and floated. In Fig. 15-8, it will take some time to shovel concrete into the custom curbed area. Thus, the finishers have tamped and floated the main body. This will be added assurance that the main slab will not dry out.

With large slabs that have walls or obstructions on two sides, you may have to float half of the slab from one open side and the other half from the other side (Fig. 15-9). Using more than three 6-foot extensions is difficult. The poles bow, which makes the back edge of the float hard to raise off the slab. Determine this factor before the pour. Make allowances to be able to reach all parts of the slab with

Fig. 15-14. Use the hand float to smooth concrete along the edge.

float were successful. The cream will be there, and that's what a finisher needs.

Waiting too long to float will make the job harder. Instead of a lot of wet cream to work with, the moisture will evaporate. This will create small holes on the surface where the cream has sunk (Fig. 15-7). That condition will require you to work the bull float back and forth many times. By working the float, you can bring up cream in one spot and push it to the area that is dry. All of the holes must be filled in. The entire surface must have a smooth texture. After the bull floating has covered and filled those

Fig. 15-15. The entire slab should look like this after the bull float and applications of the hand float along the edges.

Fig. 15-16. Use the hand float to reach areas the bull float can't.

your tools. In Fig. 15-9, the house is on the right, and a block wall is on the left. From the form below the bottom of the picture, the slab extended out 26 feet. Because three 6-feet poles only enabled the finisher to reach a portion of the slab, he had to go to the other side to complete the bull floating. The same system had to be used with fresno.

When pushing the float across the slab, hold the poles low to the ground. This raises the front of the float so that it does not dig into the concrete. When bringing the float back toward you, raise the poles in the air. That will raise the back of the float so it won't dig into the concrete. Doing this with more than an 18-foot length of poles is hard. The poles bow and will not allow you to raise them high enough to get the back of the float off the surface. To

Fig. 15-17. A hand float is perfect for filling cracks in brick.

get the poles high enough, you would have to stand on a step ladder.

Ideally, floating should go in the opposite direction as the tamp. That would make it the same direction as the screed. If you started to screed from the left side of the slab and worked toward the right, tamping should have gone from the front to the back. By operating the bull float from the left to the right, you are ensuring a flat slab. Working the tools in opposite directions is like mowing the lawn one way and then the other. Lawns end up with a checkerboard pattern, just like some of the professional football and baseball fields have. Although some groundskeepers do this to get the pattern, I'm sure it all started to ensure that every blade of grass was cut. The same holds true for concrete. By going

Fig. 15-18. Go along all edges to assure a good float job.

in opposite directions with the screed, tamp, and bull float, you are ensuring that any ridges left by any of the three tools has been wiped away.

There are, however, some times when that can't be done—for instance, if obstructions prohibit you from operating the bull float straight ahead (Fig. 15-10). In those cases, you'll have to operate the bull float the best you can. You must float all of the surface. If you can't go in the direction you want, it is all right. The added insurance of crisscrossing will not be affected, but that will not ruin the finish. The important thing to remember is to float everything; added insurance is extra.

If your job is going to take two loads, tamp and float the first batch before the second arrives (Fig. 15-11). Unloading the second truck will take time. Meantime, the first is setting up. If the second truck has to wait a few minutes, let it. The overtime will not amount to much—probably nothing. Having the first part tamped and floated will give you peace of mind and let you concentrate on the second half of the pour. If you have enough helpers, let them take care of the first load while the second is being poured (Fig. 15-12).

Staying ahead of the concrete is the name of the game. At no time do you want the concrete to dry out and get hard before you are ready. Paying close attention to the timing, starting each phase right away, and doing it right will make the job go smoothly.

HAND FLOATING

The hand float is a wood or magnesium hand trowel. You can find many uses for this tool during the pour and the first hour afterward.

While pouring and screeding, the hand float can be used to push mud into tight corners and smooth it out. It is ideal for getting cream under mudsill lips and the lips on woodsided walls. One man should have the hand float ready while screeding.

The hand float can also be used during the pour to clean off stringer tops (Fig. 15-13). It can also remove concrete from around extending poles or posts in the slab. Use it where a small bit of mud has to be moved or smoothed. It is better than using your hand because it is flat.

The biggest use for the hand float is smoothing concrete along the edges (Fig. 15-14). Sometimes the concrete is a little dry, and the bull float cannot bring up enough cream along the edge. Because the edge of the concrete dries first in most cases, it is harder for the bull float to bring up cream. The hand float can be applied with more force. Scrubbing the surface with the hand float will bring up cream. A couple of light swipes over the rough texture will

Fig. 15-19. Hand float around obstructions in the slab.

Fig. 15-20. While hand floating along the edges, use the tool to clean the tops of the forms.

smooth it out, leaving a surface just like the parts bull floated (Fig. 15-15).

Hard places to reach with the bull float can be floated with a hand float. Remember, don't walk on the slab. If you can reach a spot by walking along a step or mowing strip, go ahead and hand float the rough area (Fig. 15-16). Essentially, the hand float is an extension of the bull float. It is designed to effect the same finish while giving more versatility.

The weight and texture of the hand float will push down any stubborn rock and move concrete. By rubbing it along the surface, you can bring up extra cream to fill in rough spots and cracks in block work (Fig. 15-17).

Edges along forms and stringers don't always get a good bull floating. It is a very good idea during or after the bull float to go along all of the edges and stringers, making sure the concrete is properly floated (Fig. 15-18). Obstructions, such as manhole covers, water meter covers, and the like, should also have a good float around them. The hand float works great, especially with round obstructions (Fig. 15-19).

While checking the edges and floating them, clean the tops of the forms (Fig. 15-20). Remove excess concrete that might fall onto the slab while you are putting on the edge. This is not a mandatory operation. If other pressing jobs exist, do them first. Clean the tops of the forms if you have the time and the manpower.

SEAMS

Seams are control joints that allow the concrete to crack in them rather than another part of the slab.

Fig. 15-21. Use a straight 2 × 4 form as a guide for the seamer.

Fig. 15-22. Seam filled with cream from fresno applications.

Although they are most popular and needed for walkways, seams should be affected on long driveway slabs, too. Used with expansion joints, seams provide a valuable aid in controlling cracks.

On a walkway, a seam should be inserted every 3 to 4 feet. This is because the walkway is not very wide, and there is not much concrete to give added strength and support. On a long driveway, I suggest a seam every 10 feet with an expansion joint every 20 feet. Unlike a patio slab, the driveway will be subjected to a great deal of weight. This is a crack factor. If the concrete should crack, the chances are good that it will be controlled in one of the seams.

To place a seam, you'll need to use a 2 × 4 as a guide. For a walkway, you can use a piece that is just wider than the walk. The board will rest on forms or can be gently laid down on the concrete. With a short 2 × 4 guide, you'll have to hold it in

Fig. 15-23. Putting on a second seam with no 2 × 4 guide.

place with one hand. Putting the seam on right after the bull float will be easy because the concrete is still wet. Because of the wet surface, the 2 × 4 will slip.

On a wide slab, you'll have to place a 2 × 4 across it, too. Chances are that its own weight will hold it in place. Place the 2 × 4 guide so that one side will mark a straight line at the point you want the seam (Fig. 15-21). Rest the seamer tool against that side and push the tool across the concrete to make the seam. Seams are best effected when the concrete is wet. The wet mixture will allow the

Fig. 15-24. Seams should be placed at every corner.

indentation has already been established, and the cream will not block the tool like a rock will. When doing the second seam, go slow. Try to go in a straight line and move the tool in a smooth and steady motion. The lines that will be made by the edge of the seamer can be taken out with the fresno. Later, you'll have to hand trowel that section, and then you can be sure to wipe out any lines left over.

For walkways, it is only necessary to have a hand seamer. You'll need a walking seamer for larger slabs. It should be equipped with the same adapter as the bull float and fresno so that the same extensions can be used for all tools.

Seams must come out from any corner (Fig. 15-24). If the slab you pour surrounds an existing piece of concrete on two sides, you had better place a seam extending out from the corner across the new concrete. If not, the new concrete will crack. At every point a walkway makes a 90-degree turn,

rocks to be pushed out of the way. Therefore, if you have to install seams, do it right after the floating. Don't wait for the concrete to get hard.

After the seam has been made, remove the 2 × 4. You'll need some help because of the awkward position of the 2 × 4. Use the bull float to remove any marks made by the 2 × 4. There is no need to keep the seam perfect during the rest of the preliminary finishing. The fresno will cause cream to fill in the seam (Fig. 15-22). Although you can run the seamer after every fresno application, it only needs to be done after every other. The cream in the seam is easily finished. There will be no rocks to contend with, and the cream will be pushed out of the seam and into the surface. Once the seam is made, it is easy to maintain.

After the first installation of the seam, you will not have to use the 2 × 4 guide (Fig. 15-23). The

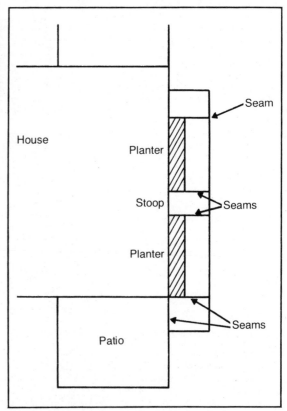

Fig. 15-25. Note the locations of the seams.

Fig. 15-26. A walking edger is easier and quicker to use than a hand edger.

Fig. 15-27. The front of the walking edger should be off the surface when walking forward.

done as soon as possible. The smooth round corner will be filled with cream, just like the seam, when you fresno. By putting it on soon, the rocks will be out of the way.

After the bull float, edges can be made. The easiest way to do them is with a walking edger (Fig. 15-26). The rounded side of the tool will be inserted between the form and the concrete. The flat part of the tool will smooth the surface, and the rounded part will make the edge. As with all concrete tools, when going forward have the front part off the concrete so it won't dig in. When bringing the tool back, have the rear up off the concrete (Fig. 15-27).

Your only guide is the form. Gently allow the side of the edger to rest against it. This should keep you going in a straight line. Watch the effected edge. When the rounded edge is made, it should be smooth. No holes or smudges should be left. If, after the first application there are imperfections along the edge, go back and edge again. Once the edge is established, it will be easy to do later after the concrete sets up.

Hand edgers also must be available. Some tight spots make it hard to use the walking edger. When you rent the tools, make sure that the arcs on both edgers match. There are different arcs available that make a ¼-inch, ⅜-inch, and ½-inch arc. The most popular is the ⅜-inch.

The hand edger is used just like the walking one. Raise the front when going ahead and raise the rear when going back. Try to apply pressure to the tool in an even way. Too much pressure on the arc side will tend to round the edge off into the side of the slab. Too much pressure on the outside of the edger will leave deep lines. An even pressure will allow the edger to make the round arc and leave hardly no line on the side (Fig. 15-28).

Lines left by edgers can be wiped away with a hand trowel. The more you have to finish on the lines, the more uneven the outside edge of the concrete will be (Fig. 15-29). If you seem to be leaving deep lines, adjust the pressure toward the arc side of the tool.

This first edge is mainly for moving rocks and establishing a curve. Later edgings will finish the concrete. The timing for this phase is not critical. It

put a seam from the inside corner out to the form. On that type of a corner, one seam will come out straight from the inside angle to the outside form on one side, and then another one from the same starting point to the other side at a 90-degree angle. Basically, you will be designing a step out of the corner piece of concrete (Fig. 15-25).

EDGES

Putting on the edges is another chore that should be

should, however, be put on soon after the bull float. Any excess cream coming out from the side of the edger can be fresnoed into the slab. You'll have problems with rock if you wait too long. Either rock will not allow the edger to create the curve, or the edger will kick up grit that will make finishing harder.

On stringer slabs, don't forget to edge along both sides of all stringers. Short stringers and those in a walkway are easily reached with the hand edger. For long ones, you will have to use the walking edger. Again, be sure you have the correct adapter so that the extensions will fit to it (Fig. 15-30). I'll show you how to effect the final edge in Chapter 16.

FIRST FRESNO

The first fresno should be applied after the majority of the sheen has left the surface of the slab (Fig. 15-31). The purpose of the fresno is to bring up moisture and further smooth the concrete. This usually should occur from 10 to 20 minutes after the bull float. Although putting it on too soon will create too much moisture and cream, it is better to do it too soon than too late. The watery sheen on the surface is a good indicator. Once it starts to go away, it is a good idea to fresno. I recommend applying the fresno right after the bull float on very hot days. Hot weather cures concrete in a hurry. Colder and damper weather allows the sheen to stay longer (Fig. 15-32). Those conditions permit you to wait up to 30 minutes before applying the fresno.

When applied at the right time, the fresno will smooth the slab considerably (Fig. 15-33). Unfortunately, as the moisture level drops, the surface will again take on a rough look. Later applications of

Fig. 15-28. Using a hand edger allows more control over the tool than a walking edger.

257

Fig. 15-29. Note the heavy lines left by the hand edger. When the concrete is wet, they will be easy to remove. Later, when the slab has set up, deep lines will be difficult to wipe out.

the fresno will smooth the slab each time. You may have to fresno three or four times. This will be covered in Chapter 16.

When applying the fresno, use caution not to trip over tools and forms behind you (Fig 15-34). Most finishers stand next to the forms and use the hand over hand method to push the fresno across the slab. To pull it back, they raise the pole and walk backward. This seems to be the easiest way. Try it any style you want as long as the finish comes out nice.

As with the bull float and other tools, you have to hold the extensions low to the ground when pushing the fresno away from you (Fig. 15-35). When pulling the fresno back toward you, raise the extensions so the rear of the fresno will not dig into the concrete (Fig. 15-36). The fresno is adjustable on the adapter. You can tilt the blade up or down. Make the adjustment before you start. You can get yourself in a bind if the blade is adjusted too high, and you can't raise the extensions high enough to get the fresno off the slab.

Fig. 15-30. A walking edger and extensions are essential with long stringers.

Fig. 15-31. Apply the first fresno after the sheen has left the surface.

The fresno is also designed to swivel on the adapter. There may be times when you are forced to use the fresno at an angle (Fig. 15-37). This is especially true when you are working in close quarters and cannot bring the extensions back straight. Using the fresno at an angle is also ideal for walkways that are 3 feet, 1 inch wide and wider. By setting the fresno at a slight angle, you can simply walk down the entire length of the walkway and apply the fresno.

MISCELLANEOUS

When working with steps, it is difficult to get a trowel under the face form to finish the space under the bottom. By making a 45-degree cut along the bottom of the face form, you can allow easy access for tools (Fig. 15-38). The cut should be made before the form is placed. By setting your power saw to the 45-degree cut, simply follow the inside edge of the wood. The piece of wood that is cut out allows room under the form for trowels. The section behind the cut is still intact and will hold back the concrete.

When waiting for the second load of concrete, rake the last edge of the poured concrete to a sloped base (Fig. 15-39). By making a sloped edge for the fresh concrete instead of a straight drop, you can reduce the chances of a cold joint. You must vigorously tamp that section, too. If the two loads of concrete do not mesh, a crack will form. Sometimes if you do your calculating homework, you can pinpoint that spot where you want to place an expansion joint. Place a temporary form at that point and finish the concrete to it. When the second load arrives, place the expansion joint felt in place and pour.

If, for example, you are going to pour a 10-yard driveway, order 5 yards on the first truck and 5 yards on the second. At the end of the first 5 yards, you can install an expansion joint of felt or even a stringer. The concrete will come to that point and stop. The second load will finish the job. To do this, you must have your calculations correct. You must also check with the concrete dispatcher to see if he will go along with it.

Removing screed pins can sometimes be difficult. Although they are designed to be pulled easily, you can pound them in so hard that they don't want to come up. The best way to loosen them is with a hammer. Smacking them on each side usually frees them. Some finishers like to use a shovel rather than a hammer (Fig. 15-40). You can reach farther out on the slab with the shovel.

Cleaning forms and tools is a chore that should not go unattended. Water applied at high pressure works great. For those stubborn bits of concrete, use a brush or foxtail broom (Fig. 15-41). All screed boards, screed forms, stakes, pins, and tools should be cleaned as soon as possible. Wet concrete is easier to remove than dry. It is nice to have one helper available to clean up. Tools that will be used again should be clean before their use. Hand trowels can be placed in a 5-gallon bucket filled with water and cleaned. Be aware of the sharp edges of steel trowels.

Fig. 15-32. The sheen will stay longer during cold and damp winter days.

Fig. 15-33. Fresno will finish the concrete to a glassy finish.

Fig. 15-34. When operating tools with long extensions, always be aware of obstacles behind you.

Fig. 15-35. When pushing the fresno away from you, lower the extensions so the front of the fresno will be off the surface.

Fig. 15-36. When pulling the fresno back, raise the extensions so the rear of the tool is off the surface.

Fig. 15-37. The fresno is designed to allow the blade to swivel.

Stirrups and J bolts should be installed during the first hour. I have described earlier when you'll need them and how to determine their positions. Using your nails and string as a guide, you can push them into the concrete. The best time to do it is right after the first bull float. Sometimes you may have to use a hammer. The wetter the concrete is, the easier it is to put them in. Be sure of their position after they are in the concrete. Recheck them after all have been installed.

You will also have to install the 2-inch wide benderboard expansion joints during the first hour if you plan to use them. Wriggle them into place right after the first float. When they are in position, use a hand float to smooth the concrete around them. It is best to cut them ½ inch to ¾ inch shorter than the width of the walkway. This will give you enough room to seesaw them into position.

When you get all of these projects done, it will be time to put on the first fresno. The first hour is a busy one. Make a list covering all of the things that must be done so that you don't forget anything.

Fig. 15-38. A 45-degree cut on the bottom corner of the step face form allows finishing tools to be applied under the form.

266

Fig. 15-39. Preparing for the second load of concrete, the first load's last edge should be raked and sloped for better adhesion with the fresh concrete.

Fig. 15-40. Using a shovel to reach out and loosen the screed pin.

Fig. 15-41. A foxtail broom or brush works well in cleaning off concrete from tools, forms, and stakes.

Chapter 16

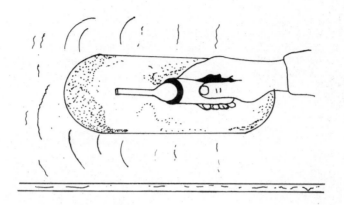

Basic Finishing

BY THE TIME YOU REACH THIS PHASE OF YOUR concrete job, many things should be done. Screeding, tamping, floating, the first seam and edge, stirrups, J bolts, and expansion joints are all phases that must be completed. The basic finishing portion of the job entails nothing more than making the surface of the slab as smooth as possible. The concrete will be setting up.

FRESNO

The first fresno application has been completed. The second application comes when the sheen has once again left the surface (Fig. 16-1). The top layer of concrete might still be wet and creamy. Because the majority of water has evaporated, it is time to fresno. Note in Fig. 16-1 the ripples on the surface. They were caused by the water leaving the slab. The fresno will smooth them out.

The entire slab must be fresnoed. The color of the surface will appear darker after the fresno application. This is due to the water brought up by the fresno action (Fig. 16-2). As long as the fresno

easily glides over the surface and brings up moisture, the concrete is still wet. You should fresno every time the sheen leaves the surface. If you wait too long, pock marks may appear. These will be slight holes created by water leaving the cream (Fig. 16-3). To fill them in, you must run the fresno across the slab several times. You may have to bring up some cream in one area and push it over to a dry spot. Work the fresno back and forth over the dry spot—putting downward pressure on the poles. The added pressure on the fresno will help to bring up the cream. After the cream has risen, apply the fresno lightly. This action will smooth the slab (Fig. 16-4).

As with the previous float and first fresno, you might have to take off an extension or two to reach a tight spot (Fig. 16-5). The purpose of the fresno is to get the concrete as smooth as possible. Low spots or cat's eyes will be areas that the fresno goes over but does not touch. The surface around them will change to a darker color; they won't. When these occur, rub the fresno over the area a couple of times. Fresno the sides of the low spot and try to

Fig. 16-1. When the sheen is gone, the second fresno will smooth the rough surface caused by moisture evaporation.

Sometimes the added weight is just enough to allow the fresno to effect one last finish. This will save you the work of hand finishing a slab that isn't quite ready for the hand trowels, but is in need of another fresno application (Fig. 16-7).

Not all slabs require the use of fresno weights. They are more of a luxury than a necessity. If available, however, you can use them to help remove low spots and bring up the last of the moisture before the hand trowel application (Fig. 16-8).

Most concrete slabs require at least three fresno applications. The timing is an important

Fig. 16-2. Notice the darker color of the concrete after the second fresno application.

get cream from around it to cover. Sometimes you can put a twisting pressure on the poles. By twisting one side of the fresno down onto the low spot, you can smooth it over. Regardless, you will have to correct the low spot by ample fresno applications (Fig. 16-6).

As the concrete sets up, the fresno will do less and less to the surface. The fresno will actually "ring" when pushed across a hard slab. At that time, you can either get out on it and begin hand troweling, or you can put weights on the fresno. Fresno weights are not uncommon. Nothing more than chunks of steel, they have a groove in them that fits over the support rib on the back of the fresno.

Fig. 16-3. Waiting too long between fresnoes will cause more pockmarks and more fresno work to smooth them.

Fig. 16-4. After scrubbing a rough area with the fresno, a second, lighter application will smooth the concrete.

factor. Although you should wait until the sheen has gone, do not wait any longer than 20 minutes in summer and 30 minutes in winter to fresno again. It is much better to apply one extra fresno than to wait too long and miss one. The fresno will make the hand trowel job easier. If many low spots and pockmarks are left behind, you will have to clean them up with the finishing trowel. All of the finishing you can do with the fresno is work that won't have to be done by hand.

Deciding when the fresno is no longer useful can be a tricky situation. The best rule-of-thumb is to watch and listen. Watch to see if the fresno is bringing up moisture and if it is doing any good on the slab. Listen for the fresno to ring. Metal rubbed against a hard surface, such as concrete, has a

Fig. 16-5. In small and tight areas, use only one extension to allow easy access.

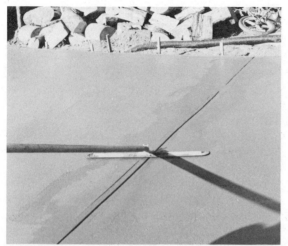

Fig. 16-6. Ample fresno work will remove a low spot on the left side of the tool. The seam can be gone over after the fresno.

definite ringing sound to it. The scraping sound is an indicator that you should begin the finishing trowel phase.

As an added test, you can start finishing with the steel trowels along an edge of the concrete you can reach from outside the forms. If the hand trowel smoothes the surface without leaving too deep a line, it is time (Fig. 16-9).

Too many finishers are afraid of ruining the slab with the fresno. For the most part, their fears are unfounded. The fresno tool is supposed to smooth and finish the concrete. Low spots, ridges, and pockmarks can be removed by the fresno. If you are having trouble with it, think about what you are trying to accomplish. Realize that downward pressure on the tool will force stubborn concrete to smooth out. Work the fresno back and forth to get

Fig. 16-7. Weights, small blocks of heavy metal on top of each side of the blade, will allow one final fresno before having to hand trowel.

Fig. 16-8. Fresno weights help to clear low spots.

274

Fig. 16-9. When you think the slab is ready to hand finish, test a section along the edge.

you otherwise would have had to get with the knee boards. Proceed along the edge around the entire slab, leaving only one open area for knee board entrance and exit (Fig. 16-11). If more than one finisher is available, have them start along another form, as well as with the knee boards.

Operating the hand trowel requires no great deal of experience. Although the finish of a professional will be smoother than that of a novice, for the most part you can do a good job if you take your time and concentrate. I highly recommend that novice concrete finishers use the pool trowel for finishing. That is the steel hand trowel with rounded corners. If you have to finish concrete that is still just a little on the wet side, the pool trowel will leave no lines from its edges. It will make the job go smoother and quicker.

The hand trowels will be swept in a circular motion, making an arc. When going to the left, the left side must be off the concrete; going to the right, the right side should be up. Holding the leading edge too high off the surface will cause a swirl

the desired effect. This advice holds true for slabs that are not too wet and yet not too dry. Wet concrete fresnoed too many times will have too much sand brought to the surface. With wet concrete, wait until the sheen is gone. Dry concrete can only be finished with hand trowels.

Use common sense. If the fresno will not do the job because the concrete has set up, get out on it with the knee boards and hand trowels. If the fresno makes an impression on most of the concrete and only one spot is giving you trouble, work on that spot. Push and pull the fresno over that spot, use downward pressure, and steal cream from a side next to the spot to cover it. If need be, operate the fresno in crisscross fashion from another side of the slab. Whatever it takes, get the concrete smooth.

HAND TROWEL FINISH

Most slabs start to set up along the edges first. For that reason, you should begin the hand trowel finish along the edge that first got concrete (Fig. 16-10). Use one trowel to lean on and the other to finish. This will allow you to reach out as far as you can to finish. Doing this will eliminate some of the area

Fig. 16-10. Start to apply the hand trowel finish on the concrete that was first poured and should be the hardest.

Fig. 16-11. Continue to finish along the entire edge.

pattern of cream under it. To find the right angle, you will just have to practice. After a few swipes, you'll get the feel for it.

Going too slow will not affect a smooth finish; too fast causes mistakes. A good solid motion is what's needed. Move your arm about as fast as you would to paint a wall with a roller brush. You can go faster or slower; it depends on you. Taking a few swipes on the slab will give you the feel. The only way to learn is by doing it.

If you have tried to finish a section of wet concrete, many lines and pockmarks will have appeared (Fig. 16-12). The surface will be very rough and unevenly textured. When you come to points

Fig. 16-13. This is a sample of rough concrete needing to be scrubbed.

like this, you will have to first scrub the concrete (Fig. 16-13).

To scrub the concrete, use the smallest trowel you have—most likely the leaner. Vigorously rub it over the concrete in a close circular pattern. This will make the cream rise to the top (Fig. 16-14). The total area that you scrubbed must look the same. The cream will fill in the holes and produce a

Fig. 16-12. Finishing concrete that is too wet will result in a rough and uneven surface.

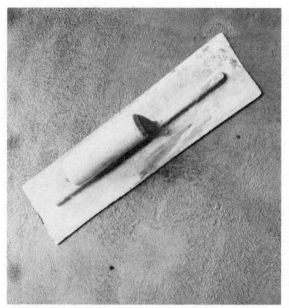

Fig. 16-14. This is how the concrete should look after scrubbing.

Fig. 16-16. Final finishing swipes should be slow, even, steady, one motion, and light applications.

texture that can be smoothed with the bigger trowel.

Only very rough spots of concrete need to be scrubbed. You should not have to scrub the entire slab. Working on a small section at a time, use the larger finishing trowel to smooth each section. Use long arcing strokes. Start at your side and continue across the front in one solid motion. The concrete

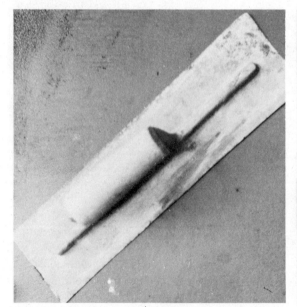

Fig. 16-15. After the scrubbing, one or two swipes with the finishing trowel will leave the surface smooth.

Fig. 16-17. Practice finishing in a remote part of the slab, preferably one that will be broomed afterward.

Fig. 16-18. You can finish small slabs from edge to edge in straight strokes.

troweled should be smooth (Fig. 16-15). The very small lines you see in Fig. 16-15 are nothing to worry about. A broom finish will erase them. If the entire slab was finished like the section in Fig. 16-15, it would be done.

Hand troweling a slab consists of two actions: first, rubbing a trowel over the surface one or two times to bring an even texture to the section; second, going over it one time in a long stroke to effect the final finish (Fig. 16-16). One of the biggest mistakes novice finishers make is working too long on one spot. The more you work the concrete, the more sand will come to the top. Try to go over an area three or four times and then put on the final trowel. If you are concerned that the finish is not good enough, go ahead and run the broom over it.

If you are concerned about your finishing ability, practice on a section before it is ready. Try to

Fig. 16-19. Trowel can go in any direction to effect finish.

pick a place that is out of the way and won't be a focal point on the concrete (Fig. 16-17). Remember that finishing wet concrete is very tough. You will leave lines. The more you try to wipe them out, the more you'll leave. This practice will help you get the feel of the trowel.

The circular motion used in finishing is not a mandatory thing. If the slab you are finishing is small, you can start at one edge and go straight to the other (Fig. 16-18). This is especially true for slabs with unique designs and small curves (Fig. 16-19). For jobs that will form narrow curbs or strips, form them so that the finishing trowel will fit between the stringer or pattern line and the outside form (Fig. 16-20).

Once all of the edges have been troweled on a

280

Fig. 16-20. Space custom stringers so trowels can fit between the stringer and the form.

small job, you can get the knee boards out on the slab and finish the center. On large jobs, you should have enough help so that at least one finisher can start in the middle when it is ready to finish. Knee boards are small pieces of plywood that will rest on the slab and allow you to walk and kneel on them. They are better to use than just walking on the slab, because they evenly disperse your weight across the surface. No deep impressions are made.

Use the knee boards in a leapfrog fashion. Set one down, step on it, and then lay down the second one. Work backward so that you can finish out the impressions made by the boards (Fig. 16-21). The goal in using knee boards is to cover the greatest amount of area by moving the boards the fewest times. Reach out as far as you can to finish the biggest area from the knee board position. Every time a knee board is placed on the concrete, there will be an impression made that you will have to remove. Sometimes the depth of the impression will require you to scrub the area first before applying the final trowel (Fig. 16-22). The more times you place the knee boards, the more impressions you'll have to finish off. Save yourself some work and strategically place the boards.

Along walls, especially rough-textured ones,

Fig. 16-21. Use the knee boards in a backward leapfrog pattern. Working in reverse lets you finish the knee board marks in front of you as you move.

you will have to operate the trowel like a putty knife. Small sections of concrete might be very rough because the fresno couldn't get close enough to smooth them. Using the tip of your trowel, pat the concrete and slide the trowel away (Fig. 16-23). If more pressure is needed, place the fingers from your other hand on the tip and push. Sometimes you can scrape a little cream off the side of the wall, excess concrete that has been pushed there by the side of the bull float, and use it to fill in. After the area has been cleaned, you can make a swipe with the trowel for the finish.

Putting on a hand trowel finish is only critical if no other finish is to be applied. The only outdoor concrete job I can think of that would not require at least a broom finish is a tennis court. You need years of experience to pour tennis courts. Therefore, don't worry about small lines left by the trowel. Try to get the slab as smooth as possible, then run the broom across it. You will be amazed at the outcome. If the hand trowel finish is any good at all, the broom lines will make the slab look like it was done by a pro.

FINAL EDGING

After every fresno, it is not a bad idea to go along

Fig. 16-22. An impression left by the knee board.

Fig. 16-23. Using the tip of the trowel to finish a rough spot next to the wall.

the edges with the edging tool. On a wet slab, you can get away with edging after every other fresno (Fig. 16-24). This will keep the edges clean and make them easier to final finish. The tool should remain flat on the slab (Fig. 16-25). Only the front should be off the slab when going ahead, and the rear should be off the slab when going back. Applying too much pressure toward the arc side will cause the edge to tilt down toward the form (Fig. 16-26). The extreme outer edge of the slab will have a slope. The edger will actually bend down that outer part and make that area fall. This has no

other defect than in looks. Too much pressure on the flat side of the edger will cause a line (Fig. 16-27). Although the line will not create any major problems, it will require a troweling to remove it (Fig. 16-28). This just means more work.

Try to maneuver the edger so that it stays flat and leaves no lines. Slight lines may appear most of the time. They can easily be troweled out. Heavy lines are harder to remove. If you constantly go over the edge and make a heavy line, there will be a permanent crease on the slab.

Rough edges caused by the fresno and evap-

Fig. 16-24. Applying an edge after a fresno. Concrete is still wet as evidenced by the sheen and wrinkles on the surface.

orating water can be made perfect. Go over the edge as many times as it takes to make that effect (Fig. 16-29). Use short strokes on rough areas. A solid back and forth motion will bring up cream and fill in any holes. After the entire section is filled, go over it one more time with a steady stroke. This will create the final edge.

The hand trowel finisher who finishes the concrete along the edges usually also uses the edger tool. Together, the finisher can finish the outer sections of the slab and effect the rounded edges at the same time. First, he should trowel out the side

of the slab, then put on the edge with the edger tool (Fig. 16-30). After the edge has been effected, he can go back and final trowel any lines left by the edger tool (Fig. 16-31). By having one finisher do both the jobs of edging and finishing the outer sides, you won't have two helpers getting in each other's way. The job will go quicker, and the outcome will be better.

Edging along stringers is a job that must not be forgotten. If the stringers are long, use a walking edger. If you can reach them with a hand edger, use it. When putting on the final edge along stringers,

Fig. 16-25. Edger should remain flat with only the front or rear parts above the surface.

Fig. 16-26. Applying too much pressure toward the arc side of the edger will lift the flat side off the surface and cause the edge to slope toward the form.

Fig. 16-27. Too much pressure on the flat side of the edger will result in deep lines on the slab.

bring your finishing trowels with you. This way, you can finish the edges and the concrete around them while you are at that location (Fig. 16-32). You will need the trowels to wipe out any lines left by the edger tool. Because you are there already, go ahead and finish the nearby concrete, too.

Expansion joints in the middle of slabs need edging, too. While on the knee boards finishing, use a hand trowel to finish the concrete next to the joint and fill in any areas that need it (Fig. 16-33). Use the edger to effect the edge next to the expansion joint (Fig. 16-34). Use the trowel to wipe out any lines left by the edger. When the edge has been effected, go ahead and finish trowel all of the other concrete you can reach. You are finishing the concrete in the middle of the slab, and you are putting on an edge next to an expansion joint section in the middle of the slab. This maneuver will require that you place the knee boards in that area only once.

Notice in Fig. 16-34 that the finisher edged both sides of the expansion joint at the same time. This will eliminate the need to put on an edge while finishing the other side of the expansion joint.

Final seams are done much the same way as edges. After the final fresno, run the seamer along the groove and effect a good seam. Don't worry about the lines made by the tool; they can be troweled out when finishing. The main point is to get a smooth seam (Fig. 16-35). Work the seamer tool back and forth along rough sections. The rounded corners must be smooth. There should be no need to use a guide, as the groove will serve as a guide itself. If you have trouble with the walking seamer, take the extensions off and use it as a hand seamer. When you go out to that area of the slab to finish, bring the seamer with you. At that time, put on the last seam and use the trowels to finish the area on each side of it.

FINISHING STEPS

Steps pose some unusual finishing problems. Because some stakes will be in the concrete to hold the step face form, holes will be left in the concrete after they are pulled. If step face forms are not pulled in time, the concrete under them will not get finished, and there will be a permanent impression

287

Fig. 16-28. After edging, you can use a finishing trowel to remove any lines. Start at the form and apply just enough pressure to remove the line. The trowel can be lifted off the surface halfway through the stroke.

in the concrete (Fig. 16-36). When using 2 × 4 step face forms, cut a 45-degree section out of the bottom of the form. Place that open part toward the outside. You can easily get the trowels in there to finish the concrete.

The timing for pulling stakes and step face forms is crucial. If you pull them too early, the concrete will fall away. If you pull them too late, finishing the step will be a real chore. Generally, the best time to pull the stakes and form is after the concrete has set up enough to be ready for a final fresno. If you have steps on your job, be sure to have enough help to take care of the rest of the slab while you work on the steps. The process will take

Fig. 16-29. Go over the edges as many times as it takes to make them smooth.

some time, and you will not want to neglect the rest of the slab.

After you remove the nails from the stakes, gently tap the stakes to loosen them. Try to remove them with the least amount of sideway movement as possible. Pull straight up to avoid disturbing more concrete than necessary (Fig. 16-37). Being

prepared for this job, you should already have some leftover concrete to use in filling the holes. Use a small masonry trowel or putty knife to fill in the holes (Fig. 16-38). You can fill in each hole as it comes or pull all the stakes first and then fill in the holes. Either way is fine.

When the holes are filled with concrete, use

Fig. 16-30. Applying the edge on a curve may require you to use only the front part of the tool.

pulled away when the form is moved. One way to avoid this problem is to coat the inside of the form with motor oil or diesel fuel. The petroleum products will prevent the concrete from sticking to the form, making form removal much easier. After the

Fig. 16-31. Short strokes from the edge are fine for removing edger lines.

the masonry trowel to smooth the surface (Fig. 16-39). Try to get the concrete as smooth as you can. Don't worry about getting it perfect; you'll have to go over the area with a regular finishing trowel later. If the step is too far to reach from outside the forms, you will have to use the knee boards (Fig. 16-40). After all of the stakes have been pulled and their holes filled and roughly finished, you can begin to pull the face form.

Use extreme caution when pulling this form. Concrete often sticks to the side of the form and is

Fig. 16-32. Use your imagination to reach hard to get stringers.

concrete has been poured into the form, you can tap on the side of the form with a hammer. This tapping action will push the rocks away from the form and cause cream to lay against it. The layer of cream next to the form will make facing the step easier and form removal safer.

Even though you might have coated the inside of the form with motor oil, don't expect all of the concrete to stay in place. If you were to just pull the form off, I'm sure a lot of cream will come with it. Therefore, use a hammer to gently tap on the form. Start to slowly pry it away from the concrete. If the form seems to stick, tap on it some more. By slowly tapping and prying the form away, the concrete will have a chance to release from the wood. If the form is more than 3 feet long, you should have someone help you. Work together to slowly remove the form and lift it away from the concrete.

Under no circumstances should you pull up on the form. Always pull it away before any lifting.

Moving it straight up will cause the concrete to come with it. That will cause the step face to fall apart, and repairing it will be a chore.

If a little concrete comes off with the form, simply scrape it off with a trowel and put it back in place. Patching the face is just like patching a small hole in the wall.

When the face is patched, use your trowel to finish off the side of the step. This will be just like finishing the slab. If the edge is not right, get a small 2 × 4 and place it against the step face. With it acting like the regular form, use the edging tool as normal. When the edge is finished, use the finishing trowel to wipe out any lines left on the face. Continue to trowel the face until you have the smoothness desired.

It is usually easier to broom the face than finish it completely. It is difficult to effect a really nice finish on the step face. The position is awkward, and most trowels are not made for that type of

Fig. 16-33. While on the knee boards, finish areas next to expansion joints and seams.

Fig. 16-34. When finishing next to expansion joints and seams, bring the edger or seamer with you to finish them. Edge along both sides of the joint.

finishing. If it is imperative to have a clean finished step face, there are some special step trowels available. To use them, you'll have to do a special job of forming.

Brooming the step face is easily done with a foxtail broom (Fig. 16-41). If the rest of the slab is to be broomed, there is no reason not to broom the step face.

Providing an exposed aggregate finish on steps is a very delicate operation. For the most part, I recommend that novice concrete finishers avoid exposing steps. The extra water used to expose the rock will weaken the concrete. Taking the step face form off too soon will allow the step to simply fall away and ruin the job. Unless you have experience exposing step faces, stay away from it.

Fig. 16-35. The last seamer application can be applied before hand trowel finishing.

If you study the procedures of finishing concrete and have a good idea what has to transpire, you shouldn't have any problems. If you have never been exposed to any concrete work before, start out small. It is never a bad idea to get advice from others, especially if they have done concrete work before. Every concrete finisher in the world has had to learn the craft from somebody else. The person you ask for advice had to ask somebody else when he first started out, so it shouldn't hurt him to help you. If you are really concerned about your ability to finish, watch some finishers in action. Then buy a couple of bags of ready-mix concrete in the bags. Use them to pourt a small stoop for the trash cans. This will give you some practice with the finishing tools and a better idea of what finishing is all about.

Fig. 16-36. A 1½-inch strip at the bottom of the top step is an impression caused by a step face form. The form should have been pulled early enough to allow concrete finishing.

Fig. 16-37. Pull stakes straight up. It will result in the least amount of damage to the surrounding concrete.

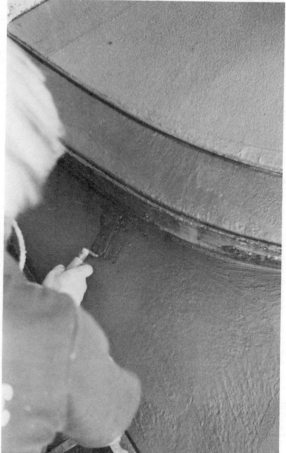

Fig. 16-38. Use a masonry trowel (pictured) or a putty knife to fill in holes from stakes.

Fig. 16-39. After stake holes are filled, use the tool to finish the surface. On colored concrete, be sure to use cream that has been mixed with the color dust.

Fig. 16-40. Use knee boards to gain easy access to the step area.

Fig. 16-41. To ensure a finished step face, use the foxtail broom.

Chapter 17

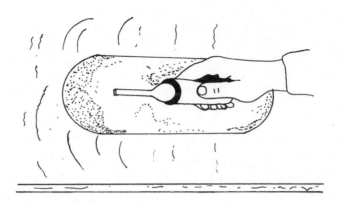

Custom Finishes

ONCERETE THAT IS LEFT WITH A SMOOTH SURface looks plain and is extremely slippery when wet. Many people have slipped at one time or another on wet concrete. A perfect example of smooth concrete is your garage floor. Because garages are protected from the weather, the slab is left smooth to make sweeping it easier. When water is applied to the floor, it is as slippery as ice.

For that reason, all outdoor concrete must have an added finish. You can apply the common broom finish or add a custom one. Some custom finishes combine both a broom and something else. The one you decide on will add traction and also give the new slab a custom appearance.

BROOM FINISH

Next to a concrete pump, a broom finish is a concrete finisher's dream come true. Besides adding traction, the broom marks cover many small finishing errors. Any slight lines left by hand trowels are wiped away by the broom. The bristles actually penetrate the surface. That action roughs

the surface and allows small bits of cream to fill in tiny holes and imperfections (Fig. 17-1).

The wetter the concrete is, the more definite the lines will be. If you needed a very rough surface on your slab, as for an inclined ramp or driveway, you can apply the broom after the last fresno. Unless the surface is not evenly finished, the broom applied that early will roughen the surface considerably (Fig. 17-2).

If the concrete has set up and is very hard, the broom will hardly make a mark at all. You will have to push and pull the broom across the surface in such cases. If that doesn't work, soak the bristles in water and then apply the broom to the concrete.

Most slabs that have been finished on time will still have enough water in them to allow a nice broom finish. The broom must only be *pulled* across the slab. Pushing it will cause the lines to be too deep. To reach the far ends of the slab from outside the forms, you need the ability to attach extensions to the broom. Earlier I showed you how to secure extensions to a wood handle. You place an extension over the broom handle and then insert a wood

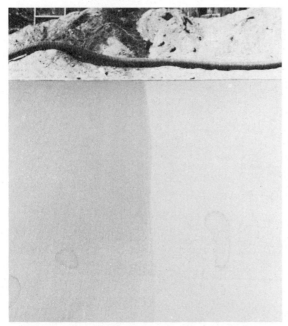

Fig. 17-1. On the left is a section of concrete broomed. The right shows finished concrete with no broom application.

use your hand that is furthest down the handle to push the bristles back and forth. To make the finish look right, you'll have to maintain the same wavy motion for each broom application.

Another way to dress up a broom finish is to edge along all the forms after you have broomed (Fig. 17-5). The edger tool will wipe out the broom lines along its path. This will create a smooth concrete border around the broomed center section. Because most people do not walk on the outer edges, this design should not hamper traction when the concrete is wet.

Some slabs have spots that are hard to reach with the extensions on the broom. For those areas, simply take the extensions off and use the broom with no handle (Fig. 17-6).

You can also apply a unique broom finish using a foxtail broom. While you are out on the slab finishing, use the short hand broom to make a windshield wiper design. To do that, you must first

screw through the hole in the extension and into the wood. If you don't have a soft-bristled push broom, you can rent a regular concrete broom at the rental yard.

Pulling the broom across the slab takes no special talent. All you do is raise the broom in the air, stretch it across the slab to the farthest point, gently lay it down, and then pull it toward you. If the impression is too deep, apply a little upward pressure on the handle to raise the bristles slightly. This will reduce the pressure applied on the slab and will result in a light brooming (Fig. 17-3).

If you want all the broom lines to run in the same direction, you'll have to maneuver yourself across the back of the slab so that the broom can be pulled straight back. On sidewalks and walkways, you can lean the broom out as far as possible. If the bristles are at an angle, it is all right. Remember to pull the broom straight back. The bristles sitting at an angle will not cause misguided lines if they are pulled straight (Fig. 17-4).

Other designs can be applied with the broom. By wriggling the broom handle as you pull, you can apply wavy lines. As you are pulling straight back,

Fig. 17-2. The center section was broomed very early, leaving a very rough finish. Edges are gone over after the broom to effect a smooth finish.

Fig. 17-3. Applying a broom finish by using extensions.

Fig. 17-4. Broom bristles are at an angle. If the finisher pulls the broom straight, the lines will be straight.

have a section of concrete finished, then extend the broom out to the farthest point you can reach. Without moving your arm, bend your wrist from side to side when applying the broom. Pull the broom straight back while doing each separate section. The end result will be a slab with a number of sections running across the slab, all in the same direction, with an equal width and equal broom arc.

If the broom makes lines too deep for your liking, simply trowel out the lines and wait for the concrete to set up a little more. Don't broom the whole slab first and then decide to change it; that's too much work. Do one section first. If the design is too deep, finish the section with a trowel and wait for a little while.

If the lines are not deep enough, put some downward pressure on the handle while you pull. If that doesn't work, try pushing the broom across a section and then pull it back over the same section. Use downward pressure as necessary. As a last

Fig. 17-5. Edging the job after the broom has been applied wipes away the broom marks from the edges. This leaves a smooth border around the slab. A heavy edger application will leave the line on the inside.

Fig. 17-6. Using the broom with no extensions to reach tight spots.

Fig. 17-7. Small patio slab with etching design.

Fig. 17-8. Try to keep etched patterns about the same size. Don't forget to round off intersecting lines.

Fig. 17-9. First, etch along all forms, stringers, expansion joints, and seams. Round off all corners.

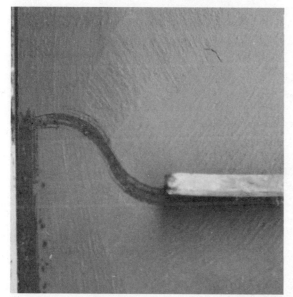

Fig. 17-10. The free-form pattern may be any design you wish.

Fig. 17-11. Holes left by rock salt granules accumulate dirt and must be washed with water to get clean.

resort, dip the bristles in a bucket of water and then push and pull the broom across the slab.

The broom is generally applied across the width on walkways. I like to broom in the direction of the slope for patio slabs—most of the time away from the house. When you wash down the patio, water will not be obstructed by the broom lines.

ETCHING FINISH

The etching finish is my favorite (Fig. 17-7). To put on this finish, the slab must be completely finished, broomed, and set up enough that knee boards will not make an impression on the surface. For a job using the etching design, you should pour the job early in the morning. You will have plenty of sunlight left when the slab is hard.

There is no specific design to follow when etching. Everything is done freehand. Make each "step" about the same size and round off every

Fig. 17-12. Spread salt heavy but evenly.

Fig. 17-13. An even application of rock salt will blend the entire slab.

intersecting point (Fig. 17-8). The depth of the etching lines should be only enough to create the solid mark. Only a slight fraction of an inch of concrete needs to be removed. I usually apply the wire brush two times—once to make the line, and the second to clean it up and ensure that all smooth parts inside the line are etched out. Some finishers like to etch the line deep enough to expose the aggregate. This is fine, but it will take more work and make the lines much deeper.

To start out, etch along the sides of all forms. Make a straight etched line along the entire edge. At the corners, round off the line (Fig. 17-9). After the slab has been outlined, get your knee boards and go out to the place where the first concrete was laid. Start there because the concrete will be the hardest. If a shady area of the slab is still wet, save it for last. You won't want the knee boards to make any impressions on the concrete.

At the starting point for the free-form etching, probably a corner, make a line from one side to the other. Use a circular design, staying about 1 foot away from the corner on each side. Round off the intersections and you are on your way (Fig. 17-10).

As you go along, you will see somewhat of a pattern starting. Try to maintain that same general pattern. Whether it be all circles or whatever, attempt to keep the pattern going. This will help give the finish a degree of continuity. Looking back at Fig. 17-8 and others in different chapters, you can see that the finishers kept the outlines about the same size and close in design. Note that all points of intersecting lines are rounded off. This gives the effect of separate stepping stones placed inside a formed area.

Again, you can make any design you want. Your only limit is your imagination. When the little bits of concrete debris from the etching marks get on the slab, don't worry. The concrete should be hard enough that they can be *lightly* swept away when you are done and completely washed away in a couple of days.

Everyone whom I have showed the etching design likes it. If you haven't seen this finish before, maybe you'll be the first one on your block to have it.

ROCK SALT FINISH

The rock salt finish puts hundreds of small holes in the surface of the concrete. Although eye appealing, the holes trap dirt and must be hosed off (Fig. 17-11). If that factor doesn't concern you, it might be the finish for you.

To make this type of finish, the concrete must be finished a little early. You will want the surface slightly wet so that the salt granules will penetrate. The top of the slab should be soft enough so that your finger will make a very slight impression, maybe just a little heavier than a fingerprint. When it is troweled to a finish, apply a very light broom finish. Use upward pressure on the broom handle to ensure the light application of the bristles.

After the light brooming, spread the salt

Fig. 17-14. You won't need to use knee boards to reach close areas.

Fig. 17-15. When pounding salt, use knee boards the same way you did during the finishing phase.

Fig. 17-16. Apply more rock salt in areas of light coverage to keep the finish even.

evenly over the entire slab. You will have to throw the salt on with your hand. Apply a heavy coat of salt (Fig. 17-12). When all the salt has been thrown on the slab, it will look like it has just snowed. Maintaining an even coverage is important. A spot that is too heavy will create an unusual pattern compared to a spot that has been salted too light (Fig. 17-13).

For areas that are not very wide, you can reach out to them without using the knee boards (Fig. 17-14). Use a hand trowel or magnesium hand float to pound in the salt. Keep the tool flat. The concrete will still be a little wet. Any side of the trowel that hits the surface will create a divet in the surface. Use as much force as necessary to get the granules embedded in the concrete.

For slabs, you will have to get out on the knee boards. Because the salt will be sitting on the surface of the slab, the bottom of the knee boards will rest on them rather than the concrete. Use the knee boards just like you did when finishing. Leapfrog around the slab and reach out as far as you can. Pound down every granule of salt within reach. When the area is done, move on (Fig. 17-15).

When all of the salt on the slab has been pounded down, you are done. If you see a section that has too light a salt application, spread more salt on it. Then go out and pound it down (Fig. 17-16). Don't wait too long. All the time you are on the slab pounding salt, the concrete is still setting up. If the concrete gets too hard, you will have a difficult time pounding the granules into the surface.

Some ingenious concrete finishers have developed a rock salt roller. It is simply a piece of 4- to 5-inch plastic pipe fitted with brackets and an extension adapter (Fig. 17-17). The inside of the pipe is filled with concrete that adds weight. At about 25

Fig. 17-17. Rock salt roller with an extension attached.

about 4 yards of surface concrete. If your job is under 4 yards, I'm sure one bag will be enough. If the job is 4 yards or more, get two bags.

EXPOSED AGGREGATE

An exposed aggregate finish is effected by washing away the top layer of cream. Under no circumstances should you tamp an exposed slab. The rock must remain close the the surface. All that is needed is a good screed job and float. The slab will be finished as any other. You will have to edge, place seams and expansion joints, fresno, and even hand trowel. The hand trowel finish does not have to be perfect. The time you will have to do it is when the fresno doesn't finish the slab satisfactorily. Essentially, the slab should look like it has been finished before you go to expose it. The reason for this is so that the aggregate will be at about the same height. You don't want the slab to have many waves across the surface. The slab should be flat, the edges round, and the surface smooth.

Fig. 17-18. Use a soft-bristled broom to loosen the top layer of cream.

to 30 pounds, the roller pushes in the rock salt. It works quite well and is inexpensive to make. The most expensive item will be the extension adapter. The roller will push down most of the salt, but possibly not all of it. If the salt rock roller doesn't do a complete job, you will still have to pound any salt that is not properly embedded.

The rock salt does not have to be removed from the slab. After two to three days, water applications to the slab will dissolve the salt. The only concern is if very salty water will run into and damage nearby flowers.

The rock salt I recommend using is Morton brand coarse. I have had good luck with the size of the granules. For the most part, one large bag of salt will cover approximately 320 square feet—

Fig. 17-19. Sweep away excess concrete and sand.

Fig. 17-20. Use moderate water spray to remove the surface layer of cream.

The slab should be hard enough to walk on before you start exposing. After it has been finished, try to walk on a section. If you leave an imprint, the concrete is not ready. Once you can walk on the slab without leaving a footprint, you can start.

Begin by sprinkling water on the slab. Be very careful not to use a straight stream of water; use a spray. The straight stream will dig into the con-

crete and make a slight hole. After water has been applied, use the soft-bristled broom to loosen the top layer of cream (Fig. 17-18). Push and pull the broom across the slab. There will be a lot of mud broken loose. You can sweep it off to the side with the broom (Fig. 17-19).

As the broom sweeps away more and more of the cream, you can use a moderate spray to remove concrete and expose the aggregate (Fig. 17-20).

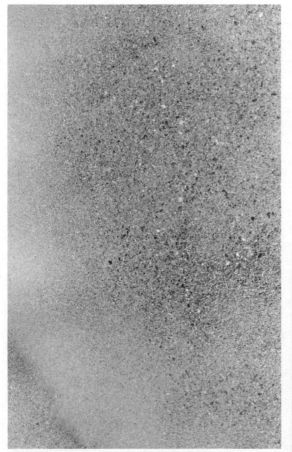

Fig. 17-21. Wash away only enough cream to expose rock.

As more of the cream is swept away, water will work better. Using a moderate spray, constantly keep the nozzle moving. Letting the spray concentrate on one spot will remove too much cream and will start to dislodge rocks. As with the broom, it is better to use two or three water applications rather than one strong application (Fig. 17-25).

If the concrete has set up so that you have to use a stiff-bristled broom, you might have to work with another helper manning the hose (Fig. 17-26). As the broom breaks loose the cream, the water can wash it away. Working together will accomplish the chore quicker and easier.

To put on a professional exposed aggregate finish, you should wash off the slab at least three

Fig. 17-22. The broom should be used to sweep away excess cream. Too much water pressure applied to an area might remove too much concrete and dislodge gravel.

You will want to wash away only enough cream to completely expose the top part of the rock. Washing too deeply will dislodge rocks and form holes. Once the depth is established, go on another spot (Fig. 17-21). As you are using the water, a cream layer will form in front of you. Use the broom to sweep away this excess cream. Using too much water and pressure to move the cream will cause you to dig into the slab too much (Fig. 17-22).

If the slab is really set up and the soft-bristled broom isn't doing much good, use a stiffer broom (Fig. 17-23). Use short, even strokes. Don't try to remove too much cream at one time. It is better to make two or three applications rather than one strong application. You will remove an even amount of cream (Fig. 17-24).

Fig. 17-23. If the concrete is set up, you can use a stiff-bristled broom.

saw another exposed job where the color was much whiter? The only difference between the two is the amount of washings. The white slab probably only had one washing. A film of cement and cement by-products has coated the surface. The darker slab is that way because that film was removed by the third wash.

The concrete needs to be thoroughly cleaned to leave a rich appearance (Fig. 17-29). Even though the job looks done, as in Fig. 17-29, go ahead and wash for the third time. You will be surprised to find that some concrete residue will still come off the slab.

When exposing a slab, wash it off as many times as it takes until the water stays clear (Fig. 17-30). To do this, the concrete must be set up.

separate times. Washing it off only once or twice will result in a grayish-white film over the top of the slab. For that reason, the first wash should mainly get most of the cream off, lightly exposing the rock. Subsequent washings will clear away more cream and sand and will expose the gravel even more.

Applying the second wash can almost be done completely with water (Fig. 17-27). More concrete will be washed away. The rock will become more pronounced, and the job will start looking better and better. Excess sand and concrete can be swept away with the soft-bristled broom. At this point, even though the concrete is well set up, a stiff brush might dig into the surface, dislodging rock. Continue to use water (Fig. 17-28).

After the second wash has been completed, you can expect to do a third. The aggregate is plainly visible by this time. You may wonder why a third washing is necessary. Have you ever seen an exposed slab that had a rich dark color to it, then

Fig. 17-24. The broom applied in short, even, and light strokes. Two or three applications are better than one heavy stroke.

315

Fig. 17-25. Apply water with a moderate spray. As with the broom, two or three applications are better than one heavy pass.

Don't start exposing until you can walk on the slab and not leave a mark. This type of finish, as well as the etching, takes time. Again, schedule an early pour. Make sure you will have enough daylight to work with. In Fig. 17-30, the third water washing took place at about 3:30 P.M.. The first load of concrete arrived at 6:30 A.M.. The day was sunny, and the temperature rose to around 75 degrees Fahrenheit. You won't have any problems during the summer. During the cold and wet winter months, the days are short and the concrete is going to take longer to set up. Be sure to pour early. You can also ask the concrete dispatcher about using a hot water concrete mix and a concrete accelerator. Those factors will help the concrete to set up faster.

If you are planning to expose a small section in the center of a slab, you will have to do the exposing a little differently. The unexposed parts of the concrete should be protected from too much water. You can place a garden hose over them to direct the water flow to a particular location. Again, the concrete should be set up enough for you to walk on it. Instead of using water pressure to remove the top layer, allow a slow trickle of water to come out the hose while you apply the broom (Fig. 17-31).

Extreme care must be used to avoid damaging any other part of the slab. Take your time and concentrate on what you are doing.

APPLYING COLOR

Color dust for a slab is available at the concrete plant. You can buy it in 1-pound bags. Generally, 1 pound will cover about 20 square feet. On large jobs, it is best to color as you go. After about 4 feet of concrete is on the ground, stop the truck and start tamping. After that, dust the concrete and then float. This will be a time-consuming job. It would be best if you had plenty of helpers.

By applying the dust as you pour, you can be

Fig. 17-26. A broom man and water man working together.

316

Fig. 17-27. Second wash will expose more aggregate and clean the surface of sand and silt.

assured of an adequate dust coverage. The wood bull float will work in the color and by the time the job is poured, it will be colored, too.

The dusting powder is very potent. It will stain anything it touches. If only part of your concrete job will be color dusted, only certain tools should be used for the colored part. If those tools become stained and are used on the uncolored part, the plain concrete will be stained. Keep the tools separated.

On smaller jobs, dust can be applied and then worked in with the hand float (Fig. 17-32). In Figs. 17-32 through 17-39, the concrete finishers are

Fig. 17-28. Most of the second wash can be done with water. If a sand buildup occurs, use the broom to sweep it away.

pouring a front porch and walkway. The step to the house is rounded and will have a curbtype border. The border will be red, and the center will be exposed. This was a very tricky job. No red dye could get on the exposed part, and exposing was a threat to the color.

To apply the dye, simply sprinkle the dust onto the area to be colored (Fig. 17-33). After a short section has been dusted, use the wood hand float to work in the dye and spread the color evenly. Keep going a few feet at a time. Try to spread the dust over a wide area with each application. This will make it easier to mix (Fig. 17-34). Using the hand float, work the concrete and color together. Rub the float across the surface in all directions; you can use circular strokes to help blend the color (Fig. 17-35).

Doing a small section at a time is time-consuming. After a portion is finished, sprinkle dust on the next area and move on (Fig. 17-36). After the entire area has been colored, treat it like any other slab. You will have to fresno, edge, and hand finish the surface. The dye changes the color.

Getting color dye on step faces is difficult. To do a good job, you will have to have some cream

mixed with the color dye in it. Use the extra colored cream to apply to the face after the face form has been removed. Because the concrete will have been sitting for a while, it will be hard. Use a steel trowel to work up some moisture out of the face, then use the trowel to apply colored cream. Finish the step face as normal.

Another way to help dust the step face is to gently pull the face form away from the concrete. It only needs to be pulled back a fraction of an inch. Insert a long trowel between the step and the form. With the trowel holding back the concrete, put some dusting powder into the area between the concrete and the form (Fig. 17-37).

After you have gone along the face form applying dust, use a small hammer or the side of the trowel to tap the form (Fig. 17-38). The tapping will cause the cream on the outside edge to vibrate. The vibration will help the cream to mix, spreading the color. This is not a perfect way to apply color to a

Fig. 17-29. Finisher starting to wash a slab for the third time.

Fig. 17-30. Wash the slab as many times as necessary until the water is clear.

Fig. 17-31. Exposing a slab on delicate jobs requires little water pressure and use of a small broom. In this photo the outer edge is red-dyed concrete to remain smooth.

Fig. 17-32. On small color jobs, use the hand float to work in dust.

step face. The color doesn't always cover the entire step. Therefore, you must still have a bucketful of colored cream to apply after the form is pulled.

As the concrete sets up, you will have to finish the colored concrete just like the rest (Fig. 17-39). While the concrete is still wet, you might want to work in the color a little more with a steel trowel. The steel trowel brings up more moisture than a hand float. The extra moisture will help to spread the color a bit more.

Colored concrete is pretty. Use only designated tools on it, don't let the dust get on anything it is not supposed to, and seal the slab to keep the dark color from fading. Different sealers are available to protect colored concrete. Some are even called waxes. For the best kind of sealer for your job, contact the concrete dispatcher. While you are at the concrete plant choosing the color, ask him about

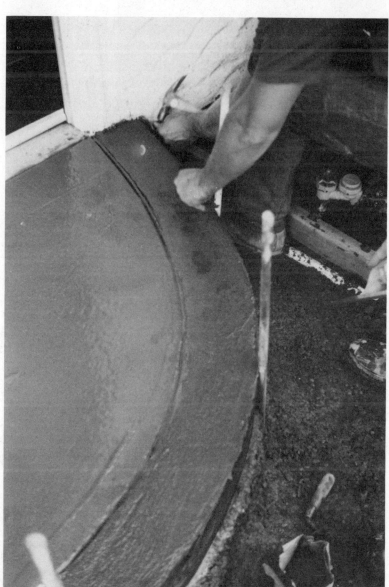

Fig. 17-33. Applying color dust by hand.

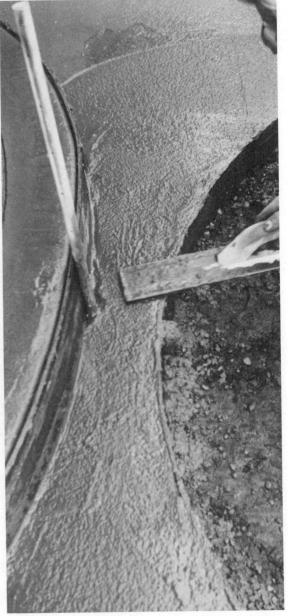

Fig. 17-34. Try to spread dust evenly over a wide area.

Fig. 17-35. Use circular strokes to blend in color dust.

322

Fig. 17-36. Color a small section and then move on.

Fig. 17-37. Pull back the form and use a trowel to separate concrete to apply color dust to the step face.

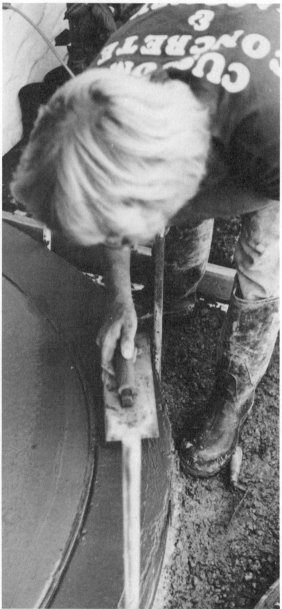

Fig. 17-38. Tapping the side of the form will vibrate concrete to help spread color dust.

Fig. 17-39. Colored concrete will be finished just like plain concrete.

Fig. 17-40. Brick insert on patio slab. Notice the slope of the slab away from the house as evidenced by the top of the slab in comparison to the level mudsill lip on the far wall.

Fig. 17-41. Gap in concrete must be wide enough for brick and mortar.

a sealer. He will either have some to sell you, or he will direct you to the place where you can get it.

BRICK INSERTS

A final way to dress up a slab is not a finish at all. Almost any kind of finish can be applied around brick inserts (Fig. 17-40). Inserting bricks into the slab takes some extra forming work. You will have to form the area as if you were making separate slabs. The area for the bricks must be wide enough to accommodate the brick and the mortar (Fig. 17-41).

When forming for the gap between the slabs, don't forget to include the width of the form. Two 2 × 4 forms measured along the top will equal 3 inches. If you allowed just the right amount of room for the bricks in between the forms after the forms were pulled, you would have an extra 3 inches added to the gap.

Any design is possible. You can figure for a straight insert, or you can make designs with the bricks (Fig. 17-42). Plenty of forethought must go into this type of job. Extra forms and stakes will have to be available, and the measurements must be accurate.

Fig. 17-42. Any brick pattern can be designed into a slab.

Fig. 17-43. Unique brick insert dresses up a slab.

326

To install the bricks, first lay down a base of sand in the gaps. Lay all of the bricks down in the pattern you want. Be sure that the gaps between each brick are the same. Sprinkle a little sand into each gap. About ½ inch of sand should be enough to keep the bricks in place. From a bag of ready-mix mortar, pour dry mortar over the bricks. Gently sweep the mortar into the gaps. When all of the gaps have been filled, you can gently sprinkle a light mist of water over them.

More than one water misting is required. Continue to mist the water until the mortar will take no more. This slow action ensures that water is getting all the way to the bottom of the mortar joint. When the mortar is totally wet, stop the water. In a short while you can begin to gently clean the bricks with a soft brush.

After the mortar has had time to set up a day or two, you can clean the bricks with a wire brush and water. Do not disrupt the mortar joints. They should be left alone.

Brick inserts are a beautiful way to dress up a custom slab (Fig. 17-43). The extra work it takes to do them will pay for itself.

Chapter 18

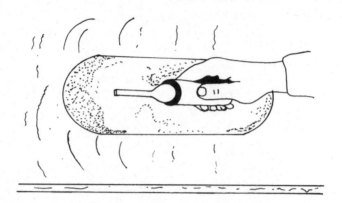

The Last Hour

CONGRATULATIONS! BY THIS TIME, YOUR CONcrete job is finished. You can sit down, relax, and have a cold drink. If you have kept up with the job so far, there shouldn't be much cleanup. Take a few minutes to rest and admire your work. Don't rest too long; there are a few chores that you must do.

CONCRETE PROTECTION

Concrete needs 99 years to reach its full strength and hardness. The majority of that strength is reached in about two weeks. A slab is very susceptible to cracks during these two weeks. The concrete must be protected from drying out too soon, especially during summer.

One way to keep the moisture in the concrete is by water application (Fig. 18-1). For two weeks, you should sprinkle the slab with water at least six times a day. Ideally, it should be wet down every time a section begins to dry. You can even build a dam around the slab and let water stand on it for two weeks.

The only drawbacks to water application are the time element and the checking feature it might create on the surface. Water, when applied to the slab, cools the concrete. When it dries, the concrete warms up. The constant changes in temperature causes thousands of very tiny hairline cracks. The cracks do not pose a strength problem; they are only cosmetic. To eliminate that possibility, you can protect the slab by two other methods.

Heavy-duty plastic can be laid over a hardened slab to help keep in the moisture (Fig. 18-2). It is best to use black plastic. The sealing action keeps the concrete's own moisture from evaporating. This is probably the cheapest method. Be certain that the concrete is fully hardened. If the concrete is still wet, wrinkles in the plastic will scratch the surface.

Plastic also is essential during the winter—not so much to keep the moisture in, but to keep out the rain and cold. Even cured slabs can be ruined by raindrops and water falling off the eaves. After the concrete is hard enough, place a piece of plastic

Fig. 18-1. If no other protection is provided, apply water to the slab at least six times a day for about two weeks. This will keep the slab from curing too fast.

over it. You must keep the concrete from freezing. To ensure that it doesn't, cover the concrete with plastic, then put a heavy coat of straw over the top. The straw will act as insulation for the concrete. It will allow the heat from the concrete to stay inside the plastic and keep the slab warm.

The third way to protect a slab is by applying a sealer (Fig. 18-3). The sealer is a liquid moisture that protects the slab by sealing the surface so no moisture can escape. It is easy to apply if you have an applicator. The sealer can be purchased at the concrete plant and comes in gallon containers. To

Fig. 18-2. Heavy-duty plastic placed over the slab will prevent moisture evaporation, keep the slab moist, and allow the concrete to set up completely.

save a couple of dollars, bring along some plastic milk jugs. The concrete plant usually buys the sealer in bulk and fills your containers. To determine how much your slab will need, contact the concrete dispatcher. Various brands have different mixing instructions.

CLEANUP

Concrete work is messy. Tools have a priority (Fig. 18-4). If concrete is allowed to cure on the tools, they will soon become useless. Use the 5-gallon bucket of water to keep hand tools clean during the pour. Afterward, you can thoroughly clean them. Forms are just as important. Screed forms and step face forms should be cleaned. If new lumber was used to form, they can be cleaned and be very usable for another project. Stakes should be cleaned, too. They can be very useful around the

house as plant supports and garden stakes. A few minutes of cleaning will keep them in usable shape.

There will be messes to clean up around the job site area. The concrete truck had to rinse the chutes somewhere. If that somewhere was in front of your house, you'll have to clean up the mess (Fig. 18-5). Although on large jobs you may be able to have the chutes rinsed at the end of the pour, most of the time they will be rinsed at the curb. The results will be sand, rock, and gray cement solution (Fig. 18-6). Good water pressure usually is enough to move the residue. If the stuff doesn't move, you'll have to use a stiff-bristled brush to get it loose.

Obstructions in the center of the concrete should be cleaned. Using an old trowel works good (Fig. 18-7). After most of the concrete has been removed, you can use a wire brush to get the rest.

Concrete that has hardened on top of existing

Fig. 18-3. Sealer applied to the concrete will prevent moisture evaporation, guard against cracks, and help the concrete to completely set up on it's own.

Fig. 18-4. It is very important to keep tools clean.

Fig. 18-5. The concrete truck driver will have to clean the chutes before he leaves the job.

concrete can be removed with an old trowel (Fig. 18-8). If new concrete is allowed to cure on the old concrete, you'll never be able to get it off. Use the trowel to scrape off most of the concrete, then use a wire brush to really finish the job.

Using water to wash off the street in front of your house is the best thing. You can even enlist the help of your kids. Use caution that they don't spray the water toward the slab (Fig. 18-9). This won't be a problem if the new concrete was poured in the backyard.

Heavy amounts of concrete can be shoveled to a dumping site (Fig. 18-10). When you are finished, the front of your house will look like it did before the concrete truck arrived. All it takes is time and energy (Fig. 18-11).

PULLING FORMS

Nine out of 10 times, forms can be pulled the same day. The only times you shouldn't pull them are on cold wet winter days and *never* when you have applied an exposed aggregate finish. Because of

Fig. 18-6. On a large two-load job, the first truck's chutes may be cleaned inside the formed area. Allow this to be done during warm weather when the water will quickly evaporate.

damp winter conditions, the concrete under the surface hasn't really had time to cure. Pulling forms the same day will expose the wet sides, and the concrete could be damaged. With exposed slabs, the large amount of water applied to them can

Fig. 18-8. Clean existing concrete with an old trowel.

Fig. 18-7. Obstructions in the middle of the concrete should be cleaned as soon as possible. Waiting too long will allow the concrete to set up and make cleaning harder.

weaken the sides. Under no circumstances should you ever pull the form on an exposed slab the same day. Wait at least two days. I made the mistake of pulling the forms on an exposed slab the same day. The results were bad. Before I left the job, a corner of the slab came off. There is still water between the form and the sides of the concrete. Therefore, the sides of the concrete are still in a weak state. By allowing the forms to stay on for a few days, you can be assured that the sides and corners will stay intact.

Pulling the forms entails nothing more than extracting the nails from the stakes, the stakes from the ground, and the form from the slab. Use a claw hammer to pull out the nails. Hit the stake with the 3-pound sledge on each side a few times. This

Fig. 18-9. The street at the front of your house will get dirty. Clean it with water, a broom and a shovel. If the fresh concrete has been poured close to that area, be careful not to direct the water toward the slab.

should loosen the stake. Try to pull the stake straight up out of the ground. If it is still in tight, hit it again with the hammer. Keep doing that until you get the stake out.

When all the stakes are removed, gently start to pull the form away from the concrete. Never pull the form straight up. If you do, some of the concrete will come with it. If the form sticks to the concrete, use your fingers to pry it away at one end. Keep working on that end until the form falls off. Pulling forms is not a job that requires a lot of experience—just caution and common sense.

Fig. 18-10. Use the shovel to remove large piles of excess concrete. The excess can be placed in a special area to be hauled to the dump later.

Fig. 18-11. When the mess is cleaned up, the entire concrete job looks better.

REPAIRING EDGES

If a small piece of edge comes off with the form, don't panic. Sometimes a sliver of wood on the form will get into the concrete. When you pull the form, the edge comes with it. To repair that, get an edger and a piece of 2 × 4. The wood only has to be a few inches long—long enough to cover the damaged area. Get the mud off the form that fell off and maybe a little cream from along the form. Use the steel trowel to get it. Roll it in your hand to make a wad of mud. You may have to add a few *drops* of water. With the 2 × 4 laid against the side of the concrete where the damaged edge is, place the wad of mud onto the damaged area. While holding the 2 × 4 with one hand, work the hand edger across the area until the edge is repaired. This process works well.

It is mainly for convenience that I try to pull the forms on the same day. If a sliver of wood is in the concrete and I don't pull the forms until the next day, I may not be able to repair the edge. If, however, I wait a few days, the concrete will be hard enough to remove the sliver from the wood. If a chunk of concrete comes off a few days after the pour, it can be replaced with concrete glue, which is available at local hardware stores and lumberyards. Follow the directions on the bottle and you won't have any problems.

Index

Index